Creating Value Through Packaging

Unlocking a New Business and Management Strategy

Jim Peters
Brian Higgins
Michael Richmond, Ph.D.

Creating Value Through Packaging

Produced by:
DEStech Publications, Inc.
439 North Duke Street
Lancaster, Pennsylvania 17602 U.S.A.

Printed in the United States of America
10 9 8 7 6 5 4 3 2 1

Main entry under title:
Creating Value Through Packaging: Unlocking a New Business and Management Strategy

ISBN: 978-1-60595-087-7

Contents

Preface

Some say we're in the "post-marketing era." Others use the term "post-consumer era." Whatever term you use, we're already into its outer edges, and it is changing business directions. Three factors dominate it:

Consumers. They drive it by having more knowledge at their fingertips than ever before. In many ways, they are better equipped than the businesses that woo them. Social networks and content from peers shape buying decisions more than advertising and marketing plans.

Global changes. The bottom of the pyramid is a huge market and serving it is the right thing to do. This global market needs business models that go well beyond the traditional ways we have marketed products.

Holistic strategy. Savvy packaged-goods businesses have to embrace new ways to deliver products that fit consumer needs. It takes a management strategy that goes all the way from the first germ of an idea to the point where the product's ultimate consumer discards the package. We call that the Holistic Packaging system.

Within that spectrum, we see packaging growing beyond being a support function that puts products in some kind of container. It becomes a strategic function because it supports new directions. It is "Big P" packaging, and here's how it has a huge impact. In the marketing era, a business could thrive by offering a good product at a fair price. By 2020, successful packaged goods companies will have to offer the best product at the lowest price. That's both businesses that sell to consumers and those who sell in the business-to-business arena. Advertising and marketing will no longer define "best product at lowest price." Consumers—and their "buying" agents, the retailers—will define that. Government and non-government agencies will more strongly influence product offerings, providing the administrative tools to implement "green" and social agendas. Packaging helps address those directions.

Business success in the new scenario will rely on a holistic strategy. The systems approach becomes mandatory for success, and companies have to do a better job of integrating all the functions involved in development, execution and delivery of a packaged product. We're going to suggest that "Big P" packaging is growing as a function to fill this need. It is a business function that spans the entire range of a packaged goods company's activities and brings the holistic agenda to the table. In doing that, it becomes a strategic enabler that helps other functions deliver benefits to the consumer.

- Like procurement, it reaches into quality assurance of suppliers, and increasingly, their suppliers.
- Like R&D, it reaches out to the technical horizon as it seeks both needs and answers.
- Like production and operations, it contributes to efficiencies that hold costs down.
- Like marketing, it ties into consumer research and into marketplace trend analysis. Packaging also communicates the brand, often reaching people in places that advertising can't. It defines what a brand or product is to potential buyers, and it does that on a global basis. It is the silent salesperson.
- Like sales, it answers customer needs, specification and idiosyncrasies.
- It embraces sustainability efforts. Sometimes it is the most visible statement of a company's commitment to "green" business practices.
- It plays a role in defining margins and profitability.

Our premise is that if Big "P" holistic packaging systems span all these activities, then management needs to see it as a strategic enabler rather than a technical support function. Those who work within packaging need to enhance packaging's strategic role by embracing the bigger picture. It is one big step toward surviving—and even thriving—in a world whose economic model is being reshaped.

Our background is in packaging and its business functions. Packaging's historic role has been to put the product in some kind of container, protect it through distribution, carry a promotional message on store shelves, and make it easy for the consumer to use the product. We have found that the process is getting more complex—at about the same pace that business is getting more complex. If the packaging function fulfills its strategic role, it can help improve product success rate, speed time to market and help business move on a more strategic path needed to thrive in a changing world.

In describing the why's and how's of doing that, we want to thank our collaborators. Something of this scope cannot be the product of just three people. As we've gained depth in our understanding of packaging, we also gained a network of insightful people who have helped us do that. Without their thoughts, this perspective would not have emerged.

JIM PETERS
BRIAN HIGGINS
MIKE RICHMOND
October 2012

1 || *Five Seconds that Signal Your Future*

A scenario . . . It is early morning in Atlanta and early evening in Mumbai. Half a world apart, two shoppers scan a store shelf and reach for a new nutritional drink brand. In the U.S., it's a liquid in a 250 ml bottle; in Mumbai it's in a single-serve pouch. Each package drew consumer attention with a shape and message that clearly said it offered more value than its competitors. Doing that requires a holistic approach to packaging. On one level, the approach deals with a common element—the nutritional value and how to protect it. But the process also deals with differences—perceptions, distribution channels, packaging suppliers, communications, regulations, environmental issues and more. By integrating all the pieces, the brand owner is able to step closer to redefining a product category and being its leader globally.

The story above says this: The right packaging for a product influences the consumer's buying choice. The packaging protects the product and tells the consumer that it answers a need; at the same time, it creates profitable market for businesses. The story is a "what if" example, but it could become reality some day if product, packaging and consumer research come together. It could mean sizeable revenue for a company that does it right. It is not fantasy. We've seen parallel innovations that have achieved similar market impact globally.

Here's a real-world story that shows just what packaging and product can do. Method Home Products delivers this benefit to consumers: It brings design and an emotional connection to home cleaning products with packaging. Consumers win because Method packaging brightens their décor. The company wins, too. Industry observers say that an initial investment of $300,000 has blossomed into a business with about $100 million-plus in sales. Method gives consumers design in the home and gives them the "green" products they want. Packaging is an integral part of the process.

PROJECTING INTO THE FUTURE

If your business sells packaged goods, you have the same potential that our hypothetical nutritional drink and Method Home Products have. You also face challenges that we have seen play out across companies ranging from start-ups to *Fortune 100* firms. Today, those opportunities and challenges exist in economies around the world, and here's a thread that runs through all of them:

> Building for the future demands you take a strategic approach to the way you develop, package and market products and brands. Some would say the old ways of doing these business functions are broken and business must let go of old ideas. Strategies for the future need to embrace the emotional, psychological and branding side of packaging and products. That is as true for luxury goods sold in New York City as it is for bottom-of-the-pyramid products sold in rural India. Wherever you are, business strategies need to take a holistic approach that embraces the physical package because it is intertwined in the eye and mind of your consumer and customer.

Some businesses do that a lot better than others. Here's a backdrop: Between 80 and 90 percent of new products fail—they are off the market in less than two years, often a lot quicker. They are lost investments in both time and money.

- P&G is a global brand marketer that institutionalized a process to cut the failure rate to about 50%. If you are a glass-half-full person, that means a 50% success rate against an average success rate of only 10% to 20% for all consumer products. P&G leverages consumer insights to develop packaging, communication and brand strategy. P&G's process has an uncanny way of connecting the dots and making packaging an integral part of its strategy.
- In China, the Future Cola brand grew to be the No. 3 soft

drink brand in just a decade, behind Pepsi and Coke. It did that with packaging similar to the Western brands, but with messaging that appealed to Chinese pride. It also leveraged a distribution strategy that targets rural China. The brand's owner, Hangzhou Wahaha Group Co., boasts more than US$ 5-billion in sales.

- Target Stores compete with Walmart by tailoring the shopping experience and using packaging to build its retail brand position. The strategy helps Target to achieve higher margins and to remain a viable competitor against the world's largest retailer.

You will see more examples as you go through this book. Some involve a little luck, but the greater proportion of win-

Above all else, 'See packaging as a lever'

When we assess how leading-edge companies maintain their marketplace edge with packaging, one concept keeps popping up. Leaders use packaging as a lever to enable the entire innovation process. It is part of a systems approach, or, in today's jargon, "holistic" thinking. Packaging helps you be aware of your internal competencies, and it shows you where gaps hinder results. It embraces every facet of innovation—customer and consumer needs, design, sustainability, sourcing, and more. If you take nothing else from this book, take away the concept that packaging as a lever—coupled with holistic thinking—leads to successful business solutions.

Figure 1.1. Packaging leverages all facets of innovation.

ners comes from companies who have a packaging innovation strategy in place. Packaging can be a route to profits, if it is done right. It extends beyond the technical and operational aspects of packaging. We believe packaging helps shape success when companies look, think and embrace it as a strategic management function.

A NEW WAY TO LOOK AT PACKAGING'S VALUE

The ability of packaging to deliver bottom line results depends on the value it delivers, often through innovation. Yet, many definitions of innovation don't incorporate a link to value. We think the connection is critical, and here's the definition we use:

Innovation means:

- Developing products and packaging with benefits that increase value to the buyer.
- Positively influencing the consumer's experience, and doing that at an acceptable cost.
- Aligning product and packaging with your organization's business needs.

A few thoughts to amplify the definition:

The term "benefits that increase value to the *buyer*" applies to consumers and to buyers in business-to-business or institutional transactions. For the consumer, we deliver value from portable juice drinks; for a business that buys industrial cleaner we can deliver value in terms of ease of use. To the operating-room nurse, value is in the security of unpacking a sterile medical device. The concept of value applies, too, to transactions throughout the packaging value web—raw material suppliers to converters, packaging manufacturers to consumer packaged goods companies. Whether it is the consumer or a customer on the receiving end, delivering value is what shapes success.

Throughout this book, we also use the term "experience." In developed economies, benefits are in terms of experience—how the consumer feels about the brand and the experience with the product and package. Beyond the consumer, each transaction along the packaging value web is built on the value and experience at the ultimate point of use.

Finally, the definition signals the need to align a product and package with the seller's business objectives and goals. Each company has different strengths and weaknesses, and that means a value analysis process can give two different answers, even when the target market is the same. That's OK as long as each company aligns its actions with its own strengths and weaknesses.

Here's one way to expressing packaging's value. It can provide a way to analyze the impact of a holistic, strategic packaging process. We call it the PTIS Packaging Value Formula, and it represents a concept. It is based on the value formula you find in business management books, except that it specifically details packaging's contribution. Our version looks this:

$$\text{Value}_{\text{buyer}} = \frac{\text{Product}_{\text{Benefit}} + \text{Package}_{\text{Benefit}} + \text{Experience}}{\text{Price}}$$

Figure 1.2. Packaging Value Formula.

Who receives the value? The value subscript in our example says "buyer." That can be a consumer, and it can be a business-to-business or an institutional buyer.

What are the benefits? Both product and packaging deliver benefits; the product's role is probably the subject of a book in itself. For sake of simplicity, we'll assume a constant product benefit as we explain the formula.

The packaging benefit, as we look at its role, is a critical part of the formula; it is a piece that acknowledges packaging's strategic role. It is critical because consumers and customers buy benefits and experiences rather then buying packaging. They buy freshness or longer shelf life; they don't buy oxygen transmission rates. This changes our thinking—it emphasizes packaging's benefit rather than how it works. It is a thought process that emerged in the mid-'90s and has grown slowly into a major direction. It is still not the predominant mindset among packaged goods companies, but it is widespread enough that packaging's role as a strategic tool has probably passed a tipping point. Here's how the emphasis on benefits plays out along the packaging value web.

Compare side-by-side

The formula works when we use it to compare alternatives where packaging changes the equation. Here's a simple example—sandpaper sold in a local hardware store. One version of the product is single sheets sold in bulk; a typical cost is $0.79 per sheet. The store also sells five sheets in a paperboard sleeve; the package sells for about $0.89 a sheet. Packaging adds the benefit of convenience to one market segment—usually the do-it-yourselfer who appreciates the way the package organizes the sandpaper and is willing to pay more per sheet. If the packaging cost is less than the increased selling price, then the marketer gains a better margin based on packaging. It certainly gets more complicated in other instances, but the basic principal remains—use packaging innovation to increase value and margins. (See the detailed example in the Appendix and References.)

Benefits versus function

It has been a long-term practice among packaging professionals to talk in terms of functions; we easily say, "The packaging material has better oxygen or water vapor barrier." In a strategic approach, we need to translate that into benefits. By saying, "The package keeps the product fresher," we focus on the benefit rather than the function. Here are some of the benefits packaging offers:

- Fresher product
- Newness
- Better taste or product performance
- Convenience and time savings
- Sustainability, from the viewpoint that the consumer has a positive reaction to a package.
- Affordable luxury
- Safety
- Cleanliness
- Availability

The benefits packaging offers also influences the value formula's *experience component;* that factor arises from the consumer's or customer's perceived benefit from a product and its packaging. It really is a measure of how well the product and package communicate benefits. The result is the "AHA!" a consumer expresses when she finds something that solves a problem she didn't know she had. The experience factor also expresses the brand equity earned in the product and packaging; advertising, packaging and product help create equity. Today it includes consumer-generated equity developed through social media.

Here's how packaging's role has traditionally been seen when viewed from the functional perspective:

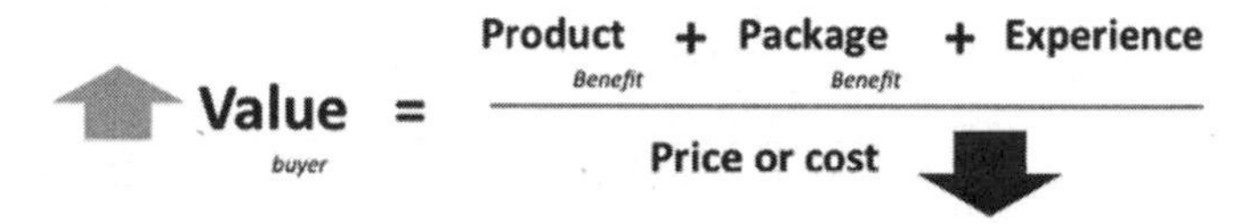

Figure 1.3. Packaging's value from a functional view.

The formula says: Cut the cost of packaging to reduce—or at least hold down an increase—in the product's price. When we look at the equation from the benefits perspective, the formula looks like this:

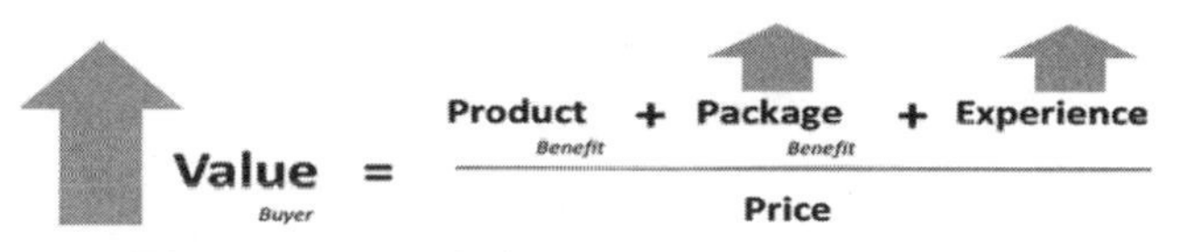

Figure 1.4. Holistic view of packaging value.

What this says is that by using packaging to deliver benefits and positively influence experience, the value to the buyer goes up. We've not indicated a direction here on price of cost of packaging, and true innovation may even allow a price decrease. The details will vary, but the increased value to the consumer can outpace the value delivered by the "squeeze the packaging cost" approach. Increasing value by innovation is what this book is about.

NINE STEPS TO DELIVER VALUE

In more than a decade of working to help hundreds of packaging companies define value, we've seen some key steps emerge.

Gut check on packaging's benefits

Sometimes, we need to take a step back and assess the benefits a package delivers to consumers. In Chapter 2, our colleague Jack Gordon offers a "gut check" management tool on how to assess if consumer insight research is on-target. Simple, he says, "Ask yourself if the benefit you think you are delivering is a *real* benefit to the consumer." Is it something that will grab the consumer's attention and make them buy what you have to offer? Will it deliver an "AHA!"? We suggest doing that for every packaging innovation. The legacy of packaging failures is littered with stories of benefits that packagers and suppliers thought they were delivering. Yet, the consumer said "ho hum" and ignored the packaged product. Asking the question honestly is a way to keep ourselves on the straight and narrow when we get caught up in our own enthusiasm.

They are not "cookie cutter" steps that apply the same way in every case. Each company has its own business objectives and processes that simply don't allow a one-size-fits-all approach. However, we've found enough common elements that build a good framework on which any packaging operation can grow and support its organization's goals. Here they are:

1. *Know the consumer and customer in depth.* Research has gotten more sophisticated. It has to, because today we need to have insights on the consumer's emotional needs. The consumer initiates the direction for the entire packaging development process. Consider the customer, too. If it is a retailer, it adds an additional layer of concern and requirements; we outline those in Chapter 3.
2. *Know your distribution network.* Packaging has to work in multiple channels that reach consumers, institutional buyers, foodservice operators and more. You need to know the paths, intricacies and idiosyncrasies of the distribution channel in which you participate. Not knowing is like going into a minefield without a map.
3. *Know and leverage the packaging value web.* We use the term "value web" instead of the more traditional "value chain" because the relationships have become so much more complex. To be successful, each company in the

value web depends on all the other organizations involved in packaging. It goes well beyond your supplier and your customer—today it includes organizations with social goals. On the supply side, you often have to follow your source of supply all the way back to the field where the crops grow or the location where a substance is mined. It is a web rather than a chain.

4. *Build effective teams.* You cannot accomplish any of the other steps without having effective, cross-function teams in place. The marketing-packaging link has become readily apparent; links to production, engineering, legal, supply chain management, logistics and others are critical, too. With these links in place, packaging becomes a hub in a network that spreads throughout an organization. In doing that, packaging truly earns its place as a strategic function.
5. *Think strategically.* Put the systems approach on steroids. It means you have to have a portfolio of projects in the pipeline. Some with immediate deadlines, some extending out five to ten years. By getting out in front, you can connect the links that deliver the best results.
6. *Build sustainability into your efforts.* In the past decade, eco-friendly, green and environmentally friendly pressures have reshaped packaging development. Sustainable thinking needs to start with the first step in investigating a new product and package, and it needs to continue through to end of life scenarios.
7. *Understand that packaging plays in a global arena.* That's why leading companies have quality control people in China. They also follow global trends on raw material, converting, technology and design trends, and they stay aware of technology being developed in other parts of the world.
8. *Think open innovation.* The days of doing it all yourself in-house are gone, even for the largest organizations. The challenge is to develop an understanding of what technology and solutions are out in the marketplace, how they fit into you business needs, and how you can work with technology owners so both sides benefit.
9. *Integrate packaging into your organization's strategy.* None of the above will deliver results unless they

are congruent with your organization's business strategy. Effective packaging executives need to know their corporate DNA. For some companies, the DNA says, "We are a high-efficiency, low-cost producer." For others it may say, "We are a leading-edge company that creates a new paths from holistic solutions." Neither is good nor bad; they are foundations on which a business is built, and development efforts need to build from the foundations' strengths.

This book's purpose is to address the details of how you can take those steps and do it with the greatest efficiency.

Checklist: Nine steps to deliver value

1. Know the consumer and customer in depth.
2. Know the distribution network.
3. Know and leverage the packaging value web.
4. Build effective teams.
5. Think strategically.
6. Build sustainability into your efforts.
7. Understand the role that packaging plays in a global arena.
8. Think open innovation.
9. Integrate packaging into your organization's strategy.

WHAT'S DRIVING THE CHANGES IN PACKAGING

The final piece of the "big picture" is the external drivers shaping the marketplace. These are the overarching events that, in the broadest way, shape our reactions to the market. One thing you'll notice is that there are a number of distinct, strong drivers that don't lend themselves to being categorized as the "new normal." If there is any way of defining a "new normal," it is that change is accelerated and the tools to deal with it also change quickly.

Driver 1. Global opportunity, with risk

We're in a global economy that is growing more slowly than in the '00s. Growth was up significantly in 2010, following recovery from the 2007 recession. It slowed in 2011 and is expected to slow further in 2012. For developed economies, pro-

	Estimated Growth, 2011	Estimated Growth, 2012
United States	1.5%	1.8%
Euro area	1.6%	1.1%
Russia	4.3%	4.1%
China	9.5%	9%
India	7.8%	7.5%
Brazil	3.8%	3.6%
Sub-Saharan Africa	5.2%	5.8%

Source: IMF, World Economic Outlook, September 2011.

Figure 1.5. Overview of world economic growth.

jected growth rates below 2% are the norm; The BRIC countries (Brazil, Russia, India and China) show higher growth rates, up to as much as 9% for China. Emerging and developing economies also see more growth, ranging upwards from 4%.

The International Monetary Fund (IMF) says two factors slow growth: First, the shift from government stimulus to private demand as a driver of growth. Second, there is uncertainty about the ability of countries to stabilize their public debt. Both are significant sources of risk. Among the other shifts, IMF sees the move from export to domestic consumption as a driver in developing economies—and for China in particular. For countries such as the United States, a move toward more exports as opposed to domestic consumption. Any moves in those directions have significant impacts on the packaging community's economic status.

Already, packagers with foresight are considering options to China. Many of China's traditional plusses are in flux. Labor costs are up, particularly in costal regions, with cost-cutting operations moving further inland. China's resolve to create domestic demand also drives labor costs up. The cost of shipping from China depends on oil prices; given the volatility in oil prices, it may be difficult to predict shipping costs.

Driver 2. The e-everything, "it's all about me" consumer reshapes demand

More and more, consumers are armed with a host of electronic communication devices. Couple that with an "it's all about me" attitude, and it fundamentally changes the way they buy. The devices and attitude work in synergy to shape a new consumer.

First, ownership and access to computers, handheld devices and mobile devices are expanding worldwide. In the U.S., for example, the number of "smart" cell phones surpassed the number of conventional cell phones in early 2012. The smart devices give consumers new services and "apps." It means that the "it's all about me consumer"—who may have been there all the time—now has a way to know more about products, brands and prices. The e-everything consumer now can comment and shape products and packaging in ways they couldn't before the electronics and social media explosion. The consumers seek out the personalized products and brands—and the experiences they bring; they also make comments on these products and brands, shaping the future development processes.

Against these trends, packaging's role is significant. With the move toward an electronic introduction to a brand, the packaging becomes the physical "portal" to the brand; that adds a new dimension to design. As the digital evolution changes the marketing environment, it also adds graphic capabilities to packages. In the simple instances, it can be limited-run package graphics tied into major events—the "big" game or a holiday. In other cases, it can be packaging that increases badge value for market segments—think of the Coors label that changes color with temperature and adds "badge" value when the young men get together for a few beers. Packaging also helps bridge the gap between mass production and the not-so-mass market. It can be the point of customization that allows a mass-produced product appear to be customized to a particular use.

A couple of added effects on the e-everything environment:

- Electronic devices alter the visual backdrop against which packaging operates. Digital technology produces richer visual context; the "temperature" of packaging graphics have to rise with the temperature of the backdrop.
- Augmented experience is another by-product of mobile technology. Consumers scan bar codes and 2D codes on packages to gain instant access to more information on the Internet, and some of that information supports experiences rather that product attributes. It makes packaging the second

most scanned object category, with newspapers and magazines being first.

A final thought on the "new" consumer. The hunt for value is as strong as ever.

Driver 3. Distribution channels are in chaos

Across the globe, retailers are morphing in ways they haven't seen before. They are fragmenting and customizing in-store and on-line experiences to meet changing buying patterns. For example, the world's largest retailer is opening Walmart Express stores across the United States, and retail analysts say it is a response to fast-growing dollar stores. Internet retailers are mainstream. Two big Internet retailers—Amazon and eBay—joined the Top 10 retailers in the U.S., according to Interbrand's annual ranking of retailers. One answer that works for retailers and brand owners is the concept of the PTIS Packaging Well Curve. (It is detailed in Chapter 3.) The concept says that those in the middle of the herd don't earn the rewards; they are earned at the edges. That's particularly true for private brands, as they become one of the largest points of competition between brand owners and retailers who are taking a more holistic approach to their packaging.

Driver 4. Green is normal

We've already said that "green" issues have reshaped packaging in the past decade. Now, the challenge is to adapt packaging management to make sustainability and "green" issues an integral part of packaging development. That means getting it into the process at the inception, and it means carrying the thinking to end-of-life. Packagers need to have a "materials strategy" and they can benefit now from looking at green science design as a tactic.

Driver 5. Design delivers a competitive edge

Design appeals to emotion and the shopper's experiential factors. Reaching those purchase decision drivers isn't just nice-to-have. More and more, packagers are beginning to understand that and give design a bigger hand in influencing branding and promotional strategy.

Designers would say that packaging reaches the "it's all about me" consumer by addressing emotional needs. Think in terms of Maslow's hierarchy of needs—the highest level is "self-actualization;" and that concept is pretty close to "it's all about me." Even at some of Maslow's lower need states, "self-esteem" and "belonging," packaging is a tool to make an emotional connection. That's true even in developing economies; Future Cola in China appeals to nationalism and a sense of ethnic pride in rural areas—certainly addressing the "belonging" need and perhaps the "self-esteem" need.

Driver 6. Science and technology offer answers

The big advances in packaging materials lie in biotechnology and nano-technology developments. Also expect information technology to drive answers. Within a three to five-year window, we expect to see some disruptive material technologies to build niches that can be the foothold for even further changes. The challenges are as often "Where do I find the science and technology" rather than "Does the technology exist." Chapter 4 contains a list of the material developments and processes to integrate them into the packaging development process.

Driver 7. It's a risky world out there

If you look at the preceding drivers, there is this thread: greater complexity and greater risk. You will find that in the marketplace, in distribution channels, among consumers and in technology development. The risk goes well beyond the traditional value chain.

- Science and technology give us answers, but they also present us with uncertainty and risk. BPA is FDA-approved, yet it has a "bad actor" reputation. Many of the risks rise from a lack of understanding of science and technology and that elevates peoples' concerns.
- Piracy, counterfeiting, terrorism and crime have direct impacts on packaging and brand equity. Consider the story of

WD-40 in China where the brand held a 50% market share and its biggest competitor was counterfeit WD-40. The problem was not only lost revenue, it was also potential for lost equity from poorly performing knock-offs.
- Empowered societies have rising expectations for safe and wholesome products. Fail to address those expectations, and your product or company could become fodder for the media's insatiable 24/7/365 appetite for news and stories. The media's goal is to build readership or website statistics, and stories with a theme of risk to the consumer build it. More than one company has suffered economically from a media "happening." In Chapter 8 we outline strategies to cope with the exposure when that happens.
- Weather and climate turbulence also create new risks and dislocations.

Smart packaging management is central to addressing all these drivers. With a strategic approach, it can contribute to delivering better value to consumers and buyers in all marketing channels. Doing that takes positive steps in developing the management process. And finally, the process also needs to be aware of the "big picture" drivers impacting industry. Even done right, it is not always going to deliver greater margins and reduce risks. But it can help stack the odds in favor of the organizations that do it right.

Key insights to the package development challenge

Know that packaging is a strategic function

Packaging Action: Packaging still contributes technical expertise that works with product development and operations, but its scope is much wider. It interacts with a wide range of business functions within an organization.

Look at the value packaging adds

Packaging Action: Drill deeply to unlock value. Often value comes from the interaction of multiple functions. One packaging format may cost more on a per-unit basis, yet it delivers savings along multiple areas of the value chain. Look, too, for packaging's unique ability to position products, extend their distribution channels and display value to consumers. Those benefits may come at a cost, so know how to show the benefit to justify the cost.

Delivering value takes a holistic approach

Packaging Action: The wider range of inputs that go into the process, the greater the likelihood that the answer will meet marketplace needs.

Know that packaging is global

It is driven by a new consumer who is armed with more mobile devices than ever before. That consumer cares about the environment, yet he wants custom products.

Packaging Action: Know that design and technology can deliver answers to those new drivers.

2 || *Look Inside the Consumer's Brain*

A scenario . . . Seattle. A woman, in her mid-20s, shops online, scanning the monitor for her favorite spaghetti sauce. She misses the package the first time; but sees the brand name the second time. Yet, the label is different. She pauses, then clicks on her favorite variety, but in a different brand. Her reaction is repeated by thousands of brand loyalists, and the brand's sales sag.

Six months earlier, a new brand manager faced eroding sales. He opted for a graphic redesign to get a sales lift. "And while we're at it," he said, "Let's get rid of the old-fashioned character on the label. We need to attract younger consumers." The designer nods and executes a new label. All without a minute of strategic consumer research on the equities within the brand's graphics. It costs the company lost sales.

Consumers who drive packaging innovation today have made a disruptive jump from where they were in 2007 when what some call "the Great Recession" began. Their attitudes have changed so significantly in the past five years that packagers, marketers and retailers may need to erase prior understandings and start with a clean slate. Those changes are significant in their impact on buying decisions and packaging that can influence those buying decisions. Knowing how to gain advantages from those changes takes an in-depth look at two aspects of the new consumer mindset: First, the "big picture," hot-button priorities and technologies are redefining consumer decision-making processes. Second, testing methods and techniques that give packagers and product marketers insights to tailor specific product and package traits to changing consumer needs.

Together, those aspects translate into a strategic effort. We've used the term Big "P" Packaging to signify the strategic approach, and it is critical in knowing the consumer. A solid strategy emerges when all the "players" in an organization—marketing, market research, packaging, design, operations, supply chain management, purchasing and other functional areas—get involved in the process. That thinking applies along the value chain, too. Consumer packaged goods companies have to be joined by retailers and by external packaging suppliers; the process needs common development goals and objectives that everyone clearly understands. Go into the process with this thinking, "Consumer testing, insight and foresight development are means to achieve an end, not go-no-go actions."

CONSUMERS: E-ENABLED, SMARTER, FRUGAL

To understand the consumer's jump, think in terms of the "e-everything" generation. The mobility and connectivity offered by electronics is changing perceptions and actions. Couple that with the "value imperative" fostered by the 2007 recession, and the consumer is buying in new ways that impact packaging.

We explored the changes with Patrick Rodmell, Principal at Rodmell & Associates, a Toronto-based consulting and creative firm that focuses on retail and leverages consumer insights into its process. The electronically enabled consumer is at the top of his list of change drivers. Rodmell uses the term "Generation M"—for mobile. It's more of a psychographic generation rather than a standard demographic; yet, it encompasses many of the Millennials. Its members were born after 1990, which means that they don't know life without the Internet. They quickly latched onto PlayStation, GameBoy and other electronic games. They have shorter attention spans, they trust social networks such as Facebook more than advertising, and mobile technology is a commodity. Most of Generation M are Millennials, yet the important fact for marketers and packagers are increasing numbers of older consumers who are adopting this psychographic pattern.

For packaging, this generation changes everything about communicating with the consumer. Perception of packaging and product attributes change when the visual is on a smart phone. The package on a store shelf projects one sense of value, but that changes when it is viewed on a screen. In looking at how

Generation M perceives packages and products, focus on these points of access to the electronic information flow:

- They rely on traditional Internet information resources—the blogs, price comparison sites, user groups, and brand owner sites. Their perception of packaged products can form even without touching a packaged product.
- They've added social media such as Twitter and Facebook as important buying resources.
- They use 2-D quick response codes on everything from newspaper ads to packaging. It enlarges their experience with packaging and product.
- They order not just from their computers, but from their smart phones as well.

A few statistics from IBM Canada confirm how deep the impact of Generation M is in that country; in the under-20 age group, more than one-third are "instrumented," to use IBM Canada's term. In comparison, among Baby Boomers, just 18 percent are instrumented. IBM Canada also sheds light on how Generation M uses Internet data. For them, comparing prices tops the list of why they go on line; accessing coupons is next. Reading product reviews ranks high. These consumers trust what their peers say;

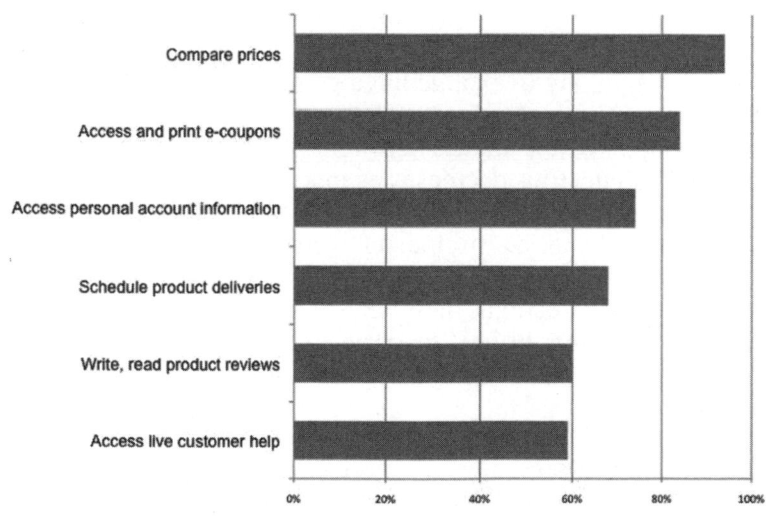

Figure 2.1. How consumers use Internet.

nine in ten people surveyed believe what other consumers say before they believe what product marketers say. Viral product reputation is a strong lever in modifying both product and packaging. Added data from Forrester Research says that on a global basis, 2.2 billion people will be on line by 2013 with Asia having the most people on line.

For packaging and marketers, the trend alters communication tactics and design strategies. The package becomes a source for extended information and active engagement. With codes read from packages, consumers can learn more and have beliefs reinforced. The e-everything generation's practices of price comparison also puts inevitable pressures on packaging costs.

What these new consumers want

If the consumer is more wired and knowledgeable, then their buying priorities should be shifting. To learn just how, we talked with Anne Bieler. She is a PTIS Associate who specializes in the consumer, retailing and retail packaging. She talks in terms of the "new frugality." It is a concept echoing among consultants and executives. Among the indicators of the new frugality—major increases in promotional circulars and coupons. Bieler notes that in 2009, coupon redemption rates increased to 27 percent, the highest rate in 20 years. Bieler sums it up this way, "The 'best value for the money' is the new shopping mantra." Consumers are willing to change buying channels to get that value; SymphonyIRI data shows that from 2009 to 2010 the number of quick shopping trips increased by almost 2 percent, while "stock up" and "fill in" trips decreased as much as 3 to 4 percent.

A supporting view comes from Kantar Worldpanel, a world leader in consumer research that covers more than 50 countries. The organization focuses on what people buy and use, and why they take the directions they take. It researched the declining market share of traditional grocery stores in England and attributed it to shoppers looking for greater value in less expensive stores and goods.

For the new consumer, conscious consumption has displaced conspicuous consumption as a priority. It's the new frugality, supported by electronics that makes price shopping possible to a degree not possible before. But it is just one part of the consumer's psyche. Here are other attributes they want in the goods they buy and how that impacts packaging.

Taste/performance

Consumers will not compromise on taste and freshness. If cost-cutting to reach a price point degrades a package's performance, it's not going to work. Consumers won't accept less than effective packaging. We'll see that unwillingness to compromise on performance echo later as we examine attitudes on sustainability.

Convenience/time savings

Time-saving products and "quick meal fixes" continue to rank high on consumer needs. U.S. prepared meal consumption is forecast to exceed US$40 billon according to Datamonitor in *Megatrends: Consumer Markets.* Growth areas reflect more demand for convenience and changes in eating patterns driven by busy, mobile lifestyles. New Product Development, Inc. reports a 30 percent growth in restaurant meals eaten at home—including everything from fast food to salads and snacks eaten as meals. Foodservice packaging takes on a bigger role in conveying a new value equation and benefits to consumers. Packaging innovation needs to deliver convenience without putting price beyond the consumers' "frugality zone."

Variety/customization

The consumer's mantra is "I want what I want when I want it. It's all about me." Social networking and mobile apps are going to support this attitude as consumers share feedback and opinions. It is going to foster changes in the ways consumers perceive value, service and experience. Consumers demand a say in what products look like and how they perform. The current perception of orderly segmentation will be replaced with new niches that won't always fit into the marketers' orderly "maps." The emerging challenge for marketers and packagers is how to interpret the insights emerging from social media and design packaging to meet them; it may even mean allowing consumers to design their own packaging to meet personal tastes.

Health, wellness and nutrition

About 90 percent of U.S. consumers believe that improving health is important, according to a study from Datamonitor. That factor is most important to those 40 years of age or older, and drives consumers to directly involve themselves in managing personal health. That trend continues to be a major driver toward preferences for all things natural and organic across age and gender groups. Packaging plays a key role here, because it can communicate the brand or product's quality and attributes. Visualize a health-conscious consumer with a smart phone running an app such as Fooducate; he's getting more than what is on the label, including the software's assessment of how the product contributes to his health.

Safety

Consumers demand assurance that foods are safe. A decade ago, consumers believed only unwrapped products were "fresh" and demanded the right to touch and smell. Today, consumers see risks associated with fresh produce and other fresh foods. The brand on a package now carries the message about reliable sourcing and product quality.

Simplicity

Shoppers want to understand easily. They want clear labeling, nutritional information, and ease of use. Regulators address that with labeling requirements, but they are not the only ones driving the bus. A number of front-of-the-package symbols have appeared, and each can have a different way of influencing the consumers' perception of the product's nutritional value. One example is Walmart's "Great for You" symbol for its private brand products. The major impact for packaging is that Walmart makes the symbol available for national branded products that meet the retailer's criteria. While it may add simplicity for the consumer, such a system adds complexity for national brands; the decision to use the Walmart brand on a product means a Walmart-specific SKU and logistic accommodations to ship packages with that symbol only to Walmart. Software also drives the need for simplicity; it may inundate the consumer with data on a product, yet software often includes a single-symbol rating. The consumer is coming to expect this as a form of communication.

Affordable luxury

The new frugality hasn't squeezed out consumer desires for small luxuries like premium chocolate and coffee. When consumers stopped going to Starbucks for expensive specialty drinks, Starbucks introduced VIA Ready Brew brand—a coffee for home or work, but at an affordable price. Individually wrapped chocolates, truffles and dark chocolate are popular, inexpensive treats. Innovative packages—such as stick-packs for coffee—help deliver an economic, yet convenient product. Package design, materials and ease of use are essential to show the value, differentiate quality and demonstrate lifestyle benefits. More often, graphics will be the lead tactic; be sure to look at emerging decorating techniques including texture inks and soft-touch coating that can add a feeling of luxury.

Sustainability and the environment

The consumer's emphasis on taking better care of the planet is one of the drivers impacting the value chain. Along with business needs, we see a push to use fewer resources and reduce waste across the value chain. In just a few years, the move to reusable bags has penetrated retail outlets. Walmart's drive to get suppliers into its Packaging Scorecard has already changed the packaging landscape. And P&G's lead in validating supplier initiatives is taking sustainability one level further back in the value chain. Consumer attitudes are a major driver of sustainable packaging and the effects are being felt throughout the value chain, all the way to end-of-life scenarios.

For all of the traits we've mentioned above, any packaging decision requires an in-depth understanding of the consumer's complex mindset. Yes, price is important. However, products that offer benefits of convenience, safety, simplicity and affordable luxury may trump price alone. They can be tie breakers in the purchasing decision. Figure 2.2 shows one way to look at the consumer's value relationship between price and benefits. The

Figure 2.2. Price versus benefits sets the value proposition.

five boxes at the top and right are strategies that offer a product differentiation with value relationships. You can give the consumer more benefits for more money and deliver differentiation. You can do the same with less for less. The key to using this equation is to assess just how much value consumers place on a specific benefit and look at the value proposition offered by competitors. In some cases, convenience may be the key benefit—in others, it may be taste or performance. It may be any of the characteristics discussed above. (See the design chapter for more information on how benefits influence consumer value perceptions, based on purchase drivers.)

CRITICAL MOMENTS TO REACH CONSUMERS

Consumers have complex demands, and we have to be just as complex as we shape packaging's communications function. There are four critical moments in the consumer's interaction with the package and product. They are the "Moments of Truth."

- The Zero Moment happens before the visit to the store or the on line shopping site.
- The First Moment occurs when the shopper encounters the package on the store shelf or sees an electronic version in an on line shopping site.
- The Second Moment is when the consumer uses the product—in the home, at work, in school or wherever the product is used.
- The Third Moment is when the shopper or consumer experiences the product and package and begins to form a repurchase opinion. It also includes the disposal and recycling experiences.

At each Moment, the package may have a role in communicating the seven attributes: convenience; health, nutrition and wellness; safety; simplicity; affordable luxury, and eco-friendliness. Packaging graphics and structure need to show congruence with the users' lifestyle. Miss one communication, and the product can join the 70% to 80% of products that fail in the marketplace.

Each Moment has is own set of defining factors. However, they are related and one provides input for the next.

Zero Moment of Truth

This is the time before the consumer encounters the packaged product. What expectations does the consumer have, based on her needs and what she has heard about the product or brand?

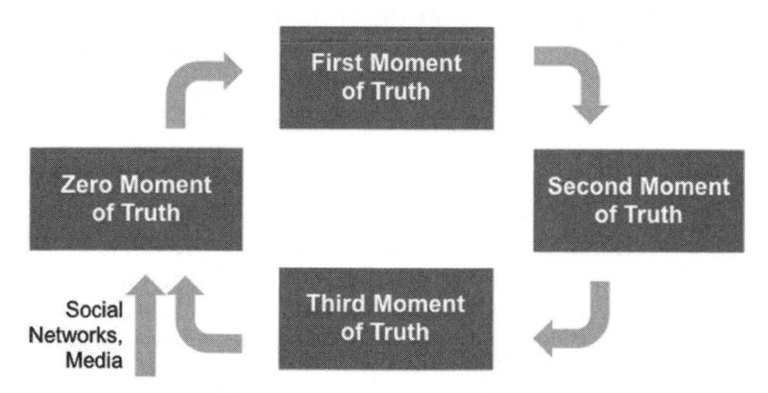

Figure 2.3. Moments of Truth.

No surprise here, but the mass of information on the Internet has become a major factor in the Zero Moment of Truth. The new shoppers head into stores—on to on line ordering sites—armed with more information than before to make product judgments. Research in 2011 says consumers are doing more research than before. It is primarily on line where they surf websites for value reviews, ratings and statistics. Often they spend at least one hour researching before making a purchase. A Google Shopper Sciences Research study says "shoppers are using an average of 10.7 different sources in their decision making process." It is at the Zero Moment of Truth that the brand or product makes it onto the shopping list. While most research says that somewhere around 70% of purchases are made at the point of purchase, no marketer wants to miss the 30% of purchases determined by what is on the shopper's list.

First Moment of Truth

The "First Moment of Truth" is the instant when the shopper first encounters the packaged product. The term gained credence in the early 2000s, and P&G helped give it that when they put together a staff whose primary job was to know what the shopper was thinking at that instant. The staff's function was also to find ways to influence thinking and actions at that instance.

The thinking of a decade ago said First Moment of Truth is the instance when the consumer meets the packaged product on the *store shelf* and makes a decision to buy. The reaction is often subconscious and shopper research estimates it is a 1.5- to 6-second window in which the consumer decides to put the package in the cart or leave it on the shelf. But, here's an important perspective for the future: That moment may be on a pad computer or a handheld device, significantly changing the dynamic.

In the store, the First Moment of Truth primarily involves the shopper's sensory reaction to design elements—graphic and the package's structure. It is color, shape, texture, and even, in rare cases, blinking lights. The chapter on packaging design will give you insights to the impact of specific elements. The Internet changes that perception, and price comparison becomes significant at the first moment. Web sites emphasize price and the visual impact of the package diminishes; one challenge in design is to project value in an environment that diminishes that component and emphasizes price alone.

Here's a way for you to get a real sense of what that means in the store: Spend some time in a supermarket high-end candy aisle. You can see just what a range of packaging color, shape and formats help define a brand and create the impact that lures the consumer's attention to the brand. Among the formats you'll find are cartons and pouches. It is a case where high-impact packaging graphics actually expanded a marketing niche by making it easier to position a candy as "high end" within the product category. In the candy section, cartons aren't just rectangles. They have curved panels at corners, an angled top that looks like a gable-roof. Stand up pouches orient the graphics, and high-gloss films add shine that grabs the consumers' attention. Brands such as Toblerone use ownable designs to differentiate the brand. These are just some of the classic tactics in high-end candy packaging.

The research at the First Moment of Truth is a unique topic that looks at the "shopper" mindset, when the hustle and bustle of the store influence purchase decisions. The emergence of "retail-ready" packaging is a new influence at that moment. The idea of using trays and shippers to simplify stocking adds graphic demands on the shipper—that could be multicolor printing and other decorating techniques. We'll go into more detail on the effects of retail-ready packaging in the chapter on the retail environment.

Second Moment of Truth

Here's where the consumer actually uses the package and product. Is the package hard to open? Does it fit into the consumer's hand easily? Does it easily dispense the product inside? Researchers say the experiences help create an emotional association with a brand. The moment of use is a focal point in determining the consumer's emotional connection with the product and package.

Here's a case story of just how the package impacts the consumer at the Second Moment of Truth. A well-known cookie maker had used a bag for years. Yet, the marketing staff was intrigued by the apparent advantage of a carton's "billboard" effect; the theory was that with more space for graphics and the printing options available on a carton, greater shelf impact would translate into more sales. But sales fell, and it was because of what happened at the Second Moment of Truth.

The brand owner invested in some ethnographic research—observing what the consumer did in the home with the package. What it found was that the original cookie bag was an inventory control mechanism. As the shopper headed out the door for the grocery store, she did a quick visual check on the cookie bag. If it was rolled down considerably, that was a "flag" to restock. The carton offered no such indicator. It could be full, or it could have just one cookie inside, but it always looked the same. The only way the shopper knew the box was empty was when someone had eaten the last cookie and threw away the carton. The out-of-stock event in the home cut down the number of cookies eaten and put a dent in sales.

On a broader front, the idea of in-home stocking opens other packaging alternatives. For paperboard, we've seen open slits in cartons that hold multiple units of product; they give the consumer a gauge on how much product remains.

The Third Moment of Truth

This is the repurchase decision. It embodies the product and package's ability to address emotional needs. It also involves the package's graphics and shape because they help the consumer find the product when they go back to the store or find it on line.

We're going to talk about consumer values on "green" or "eco" issues in a later chapter, but it is safe to point out that these perceptions fall within the third moment of truth. If the package is hard to recycle or creates a lot of waste, consumers increasingly form a negative perception that can impact the next "First Moment of Truth."

HOW THE CONSUMER SEES *YOUR* PACKAGE

The Moments of Truth and overall consumer needs give us broad guidelines on how to develop and design packaging. However, commercial success requires that we understand how the consumer responds to each *specific* package in a *specific* context. With the complexity of consumer preferences, the need for qualitative and quantitative testing of packages has never been more important. Focus groups and one-on-one interviews are workhorse testing methods, yet technology is bringing newer methods that offer deeper insights into consumer attitudes and perceptions. These insights can determine marketplace success or failure.

Neurological testing

Neurological research, in simplest terms, measure brain activity as it responds to an event. It may be the closest we've come yet to actually looking inside the consumer's head. It gives researchers insights into "what turns people on." That's the simple version. It gets more complex and sexier.

We need to take a short technology side trip to appreciate how neurological testing works. First, there's the sensing device. Basically, it is a stocking cap with sensors that fits on the consumer's head. If you're into the details, it is called a "wireless electroencephalogram (EEG) scanner." People wear it on their head, and researchers can record their brain activity. There are different versions of the cap and ways to capture data on laptops or smart phones.

The technique's value is in what it lets researchers measure. It tracks the brain's subconscious activities in terms of emotions, memory and attention by measuring specific areas within the brain. Researchers collect those brain waves. With proprietary software, researchers can map the connection between brain wave reactions and a specific event.

Neurological research isn't limited to packaging—it captures reactions to products, ads and other communications. Here's a case story that shows just what kinds of emotional connections it uncovers. The work involves a salty snack. Researchers discovered that the sticky orange-colored dust that adheres to the fingers of people who eat one snack variety triggered enjoyment. That insight led to an ad campaign to leverage the emotional reaction. If you carried that strategy out to packaging, a packaging development objective might be to enhance the sense of the orange-colored dust.

In using neurological research within the product and package development process, we see three implications:

- First, what we learn from neuroresearch usually feeds into integrated marketing communication programs where packaging joins advertising, electronic media, point-of-purchase and other media to deliver a message. Images and information have to be coordinated; often, the package is the linchpin that sets a direction for the entire program.
- Second, neuroresearch is best used today as a way to validate other research techniques. We don't see it supplanting focus groups or any of the interview techniques. At some point, its sophistication may make it a stand-alone tactic, but today it is another tool to improve the failure rate for new products and packages.
- Guiding package design where the research can give designers better data on the impact of images and icons, font structure and spatial arrangements. It delivers insights on shape, texture and product reveal.

We've already seen benefits of neurological testing in product and package development. Companies have used it to rescue failing products by confirming new perceptions of what drives consumer choice. That's true both in store and on line where this research helps us learn how to capture and hold the customer's attention.

Neurological research can be used for both shopper and consumer research. For in-store work, shoppers wear the caps as they maneuver through real or simulated store environments. And for consumer research, they are worn as the consumer interacts with the product and package.

Ethnographic Research

One of the wonderful things about being human is our ability to turn simple concepts into complex phrases. Here is one of our favorites: "Ethnographic observational research." It simply means watching what people do and asking them why. It is probably the most popular form of insight research. Consider a simple ethnographic project in a consumer's laundry room. The observer watches the way the consumer does the laundry and then asks why she is doing things in a certain way. That builds understanding. We might see the consumer carefully measure out a liquid detergent and assume it has to do with being cost conscious. But asking why gives us the insight that the consumer is trying not to use too much because she thinks it is eco-friendly. That insight that can reflect back onto packaging copy and even on a packaging's structure that allows more careful measurement. By digging underneath, we gain insight into the 'whys' to create opportunities for new products.

One of our colleagues, Brian Wagner, is a strong proponent of ethnographic research as an important tool in package development. He believes that, compared to other techniques, observing consumers is most likely to result in completely new and unexpected learning. He also points out that observation should be augmented by methods that capture products' emotional associations with the consumer. This kind of research taps into the unarticulated needs, underlying emotions, expectations, and experiences that are the basis for insight and new vision. They also discovered the importance and power of aroma and non-visual senses.

Ethnography needs to observe behavior in context – where people live, work, shop, eat, and play. The packaging research for Wrigley's "5" gum demonstrates how ethnography can guide package discovery. A group of scientists, engineers, and marketers spent time with teens and young adults—the target demographic. The observers discovered, among other findings, that the image-conscious gum-chewers wanted a resealable package with style that would look good on the table next to their iPods.

Ethnography can be used for both consumer and shopper research. For shoppers, a typical technique is the "shop-along," (researchers walking with shoppers and probing on why they do things the way they do). Mobile phones can be used to track in-store shopper experience. Used to probe consumer actions, in-home interviews and possible video documentation of the product/package in everyday use are very informative.

Social media research

The entire range of social media—blogs, Twitter, Flickr, Facebook—invites interaction and collective authorship, and it is emerging as a research tool. It is qualitative research and has its own set of interpretations and analyses. Some of the routes to gathering information can be used alone or in concert. They include:

- Polling of individuals in a community such as followers of a particular product, brand or manufacturer. Package design is a characteristic that can be researched with the understanding that the response universe may be focused on current loyal buyers.
- Polling of affinity groups such as gamers on a new package design. You may be able to get beyond you own loyal buyers with this kind of research.
- Analysis of on line generated comments related to packaging. The results are open-ended and may provide insights that can point to a number of packaging changes. Price is a key topic, but concerns on sustainability, "premium" positioning, and confusion on varieties or among brands can emerge. A caveat here is that the impact of comments may be disproportionate to the numbers. Tropicana acceded to the loud impact of those who didn't like the 2009 packaging design. Marketers for the Eight O'clock coffee brand, on the other hand, kept their 2010 redesign despite some strong negative social media comments; their belief was that their own social media research was an accurate measure of consumer acceptance.

Scott Young, president of Perception Research Services, offers these thoughts on using social media as a research tool:

> "Social media can be a good way to connect with 'fans'—your very loyal customers. You can use it as a forum for 'pre-screening' ideas such as new packaging systems and ensuring that there aren't unintended consequences such as negative reactions and confusion," Young explains. "In addition, it can be a good forum for allowing these people to connect even further with the brand, perhaps via brainstorming, co-creation and personalization options. You can use social media to reach very specific target audiences or communities, such as people facing a particular medical condition or other chal-

> lenge. Thus, for some products, it can be a great—potentially cost-effective—way to create a dialogue with a specific community that may have strong opinions and greatly influence success."

Yet, remember that it has limitations, Young continues. By definition, social media networks are very self-selecting. They tend to contain people who are more engaged with a certain product, service or issue; reactions may tend to be overstated. They're not generally representative of the larger universe of shoppers or consumers. You run the risk of speaking with the three people who would actually notice and truly care about a particular pack change!

Because of limitations such as screen size and definition, web-based studies are generally not the right place to show packaging on-shelf and speak with shoppers about shelf-based performance. In the same way, the lack of physical packs or comps limits the way to gauge functionality. So as always, it is important to ask the right questions and not invite misleading feedback. Trying to gauge comprehension, uncover confusion, or identify potential resistance generally makes sense. Trying to gauge shelf visibility or shopability—or get at functionality issues—is typically misguided. Look at social media as a good starting point or a final "disaster check." It is not a substitute for a full on-shelf evaluation with a larger, more representative sample of shoppers and consumers.

Simulated in-store research

Computer-generated store simulations for market research are becoming more popular for brand owners and retailers. While they were developed to help retailers develop merchandising tactics, they can now address how shoppers react to packages within the store. Virtual store tests can provide a more faithful representation of shelf set and retail environment. Three-dimensional environments provide a much greater degree of realism. Quantitative tools include use patterns and timing how long they stop to look at a particular package.

Once a virtual store site is established, very sophisticated test protocols can be conducted at much lower costs. For brand owners, there are good applications for package development. An

Who is a consumer? Who is a shopper?

The terms "shopper" and "consumer" are often used interchangeably, and at many levels they are. Yet, we believe that understanding each term discretely delivers better insights on which to base packaging decisions.

The "consumer" is a person who makes a product buying decision, based on a number of factors. One of those factors may be input from the product or package user; think of a mom buying a feminine care product for a daughter who will form her own opinion on the product/package's value. PTIS Product Formula says those factors encompass the product, the package, brand equity, experience, and sustainability components and services. The "consumer" brings all those factors into the store. Yet, when the consumer enters the store, they become "shoppers," subject to another set of influences that shape their buying decision. Here's how powerful those in-store impacts can be. Most experts say that about 70% of buying decisions are ultimately made in the store. The shoppers' response to package design and shelf impact become paramount in driving sales. So are in-store promotion and the overall store environment.

In terms of the Moments of Truth, see the consumer at the Zero, Second and Third Moments. See the shopper at the First Moment of Truth.

almost limitless refinement of options can be configured and screened effectively to a very high confidence level. Market researchers have said that they can confirm the results from Virtual Store Simulations with very small field tests. Speed to market is greatly improved and is gaining acceptance as the technology becomes more widely available. For brand marketers like Kellogg, Johnson & Johnson, Kimberly Clark, Heinz, and others, Virtual Retail Simulation testing provides a new level of information about shelf visibility and engagement of their packaging.

AVOID COSTLY TESTING PITFALLS

A primary goal of package testing is to give package and product developers insights on how packaging can address consumer needs. It feeds the packaging innovation process with "targets" at which technology can aim.

In a decade of managing those efforts, we at PTIS have seen ways that consumer research on packages is "bent" to the point that it is ineffective. These are pitfalls that more than one organization has stepped into; they should raise "flags" when they occur. In his decade of managing research projects, our colleague Brian Wagner has developed a short-list of pitfalls and missed opportunities that keep repeating. Topping Wagner's list of pitfalls is the way some organizations probe consumers for packaging attributes.

Test products, not packages

"When you want to test package attributes, you can't simply ask people what they think about packaging and expect to get valid answers. When asked that way, respondents try to become packaging experts, " Wagner explains. "You need them to respond to what they do know—their own thinking as they make buying decisions for products and brands." In research, it means you shouldn't directly research consumer attitudes on packaging. Rather, get their reactions to products and brands and interpret that into packaging-related insights. Wagner says the best way to do that is specialized insight groups that probe consumer perceptions on actual products and—indirectly—packaging. "Experienced researchers are continually amazed when we push them to follow this advice. They are struck by how difficult it is for the consumer to separate packaging from the integrated proposition when it is done well," he elaborates.

It works this way. As group members are recruited, they are told that they need to go out and buy actual products that they perceive as having specific attributes. They are not told to find *packages* that have those attributes; they are told to buy packaged *products*. Group participants buy paired sets of items. For a food product, they may be asked to pick products they see on a spectrum of "natural" vs. "artificial." In other cases, it may be "premium" vs. "cheap" products; or they may focus on "outdated" vs. "modern" products. The dimensions and attributes are selected from common brand drivers, and then their opposites are identified.

Panel group respondents bring those packaged products to a session. The moderator leads the discussion on both ends of the attributes spectrum, talking in terms of the product and probing why consumers purchase the items they did. "Why do you see this product as natural?" "Why do you see this product as artificial?" At the technique's core is an understanding of sensation transferrence and the fact that packaging confers attributes to the products. In observing the group, look for the packaging-related observations. Probes of "premium" products have produced comments such as, "The product in the black package is premium," or the "specially shaped bottle is more premium than the ordinary shape." Or, "the product sounds and feels crinkly and is not very special."

Wagner adds that it is not just *what* is said at an insight group, but *how* it is said. Non-verbal cues reveal true perceptions and insights. It's the "yuk" facial expression that really carries weight. To get those insights into the packaging development process, bring the technical and marketing people to observe the group, he says. Some of the observers may need training on how to watch consumers in such a group. If designers and engineers can see photos of the product groups within a dimension, they can gain tremendous insights.

The findings are a "roadmap" of consumer thinking on packaging's contribution to the product perception. With the right probing, you begin to see trends and common perceptions. One perception that has worked its way into packaging—consumers see products in foil packages as being "colder," "hotter," or more premium than non-foil packages. We know that foil will attract the consumer's attention on the shelf.

Phrase questions in terms of the product. As a project manager, preview the work of screeners and discussion guides. If they plan to ask questions about packaging, raise a red flag. Make the question, "Is the product easy to use?" rather than, "Is the package easy to use." Questions phrased that way will lead you to ample insights into the package. If you ask about packaging, you make respondents "packaging experts"; from their new-found "expertise" they will offer answers, but unfortunately they will take you off-track.

Wagner goes on to cover a "short list" of other packaging research pitfalls:

Don't start with the goal of proving a premise. Get insights from the consumer first. Probe them for their needs, expectations, imagery and emotions around a product. Then develop packaging solutions around those insights. That sounds simple, but we've seen failiures where the research goal was to prove that an existing solution met a consumer need. Those failures have included award-winning technologies that failed to address consumer needs and failed in the marketplace.

Bring the professionals involved in a project behind the glass. Each packaging project enlists different disciplines—packaging professionals, industrial designers, graphic designers, packaging suppliers, engineers and others. Each discipline sees and hears what consumers do and say differently. By getting each individual's interpretation from a first-hand observation of research participants, you get a greater depth of solutions and design elements. It adds a dimension that isn't there if only analytical researchers leading the work develop a report full of words, numbers, charts, and graphs to creative, image-sensitive designers.

Test commercial samples. Ideally, use branded packages. If you use samples that aren't of commercial quality, you will influence perceptions in a negative direction. The concept of sensation transference says that the package influences perception of the product. Consumers see prototypes as just that—prototypes. The reactions don't translate to the final packaged product. They lead to unfairly inferior results. When testing unbranded, commercial quality packages, recognize that a strong brand will likely bring even better results later on.

Focus groups do not produce data. They deliver qualitative direction, but they do not deliver quantitative information that will indicate market performance. Yet, it is not uncommon to hear people who have been behind the glass at a focus group report that "eight of the twelve people really liked this concept." Insight groups have a level of relevance, but you have to make sure to include hands-on stimuli and recognize that the consumer is not in a real world setting. Insight group preparation often includes shopping and product use exercises on the way to the session.

Never test in the boardroom. Testing in the boardroom is not in context, does not provide insights from the target audience, and provides little correlation with actual results. Yet, the practice is common even though it violates color science and design principles. It also ignores issues of store lighting and shelf location, and array within the competitive set. One real downside—Wagner reports that testing in the boardroom has gotten people fired.

Is the insight the real thing?

The result of qualitative research should be valid insights into consumer thinking. Before accepting an insight generated in research as a call for action, ask this question: Is the insight valid? Packaging's "junk yard" is full of insights that weren't. We talked with researcher Jack Gordon, CEO of AcuPOLL Research, Inc. Jack has spent more than 20 years doing the kinds of research we've been talking about, and he's come up with guidelines to tell if research is valid and capable of taking you down the right path. Jack suggests you ask yourself a series of questions about insights you've uncovered to see if they have commercial value.

- Is the Insight relevant to the consumer? Said another way: Does it help the consumer or their families fulfill a need? If not, it's not relevant. Is it something the consumer wants addressed? Is it something that is not happening in the market place now? If the answers aren't "yes," chances are the insight won't drive a commercial success.
- Does the Insight ring true with the consumer? Is the problem you think you've discovered real to the consumer? Consumers must perceive it as a problem and believe your answer is on the right track in addressing the problem.
- Is the Insight interesting to the consumer? Do they ask questions on how it works? If they are not interested in learning more, your solution will fall on deaf ears.
- Does the Insight reflect a deeper understanding of the consumer? A few following questions can help you make that decision: Is the Insight new and unique? Does it provoke an "aha!"? It needs to be an astonishing disclosure, a big step forward that has not been addressed before.
- Is the Insight dramatic? Your goal in the marketplace is to gain attention, and you can't do that unless the solution really stands out. The more dramatic the solution, the more

easily you can gain attention. GoGurt was a dramatic enough solution that it created an entirely new product category.

- Finally, Jack says, valid insights that offer commercial opportunities can be executed in more than one way. Play out the alternatives and test them; one may do a better job of addressing the insight than another.

We are proponents of insight research as a way to drive Big "P" Packaging innovation; it opens discontinuous thinking and can deliver breakthrough thinking that provides differentiation that sells. Only a few companies use this kind of insight research for packaging, and we believe the practice should be used by more companies.

SENSATION TRANSFERENCE DEFINES A PRODUCT

The driver behind packaging research is the principle of sensation transference. It says that the package is capable of conferring attributes onto a product. Done well, it confers positive attributes. Done poorly, it can doom a product.

Louis Cheskin set the groundwork for sensation transference in the 1950s. Not only did he have the acuity to notice that perceptions of products or advertising were directly related to design aesthetics. He also was articulate in relating the impact of those observations to business people making product and packaging decisions. He investigated how consumer perceptions were based on deep-seated emotional responses. These judgments are embedded in memory—associations tied to personal experience then triggered by visual imagery.

Those precepts are just as valuable today. Even high-tech methods such as neurological research support the basic premise, adding new dimensions to answers with quantifiable yardsticks. Here are the three factors Cheskin saw as drivers of purchase decisions:

- When consumers assess something they might buy, they transfer sensations or impressions they have about the packaging to the product itself.
- Consumer perceptions are based on deep-seated emotional responses.

- This critical association must be well understood and integrated for any new package to be successful.

What time has shown is that preferences and associations are dynamic and change with time, culture and product category. In recent history, black was a "no-no" for food packaging, but Breyer's Ice Cream brought out its premium offering in black and redefined the category. In 2012, black is seen as a premium color for food products in the United States. Purple became a "women's color" in the U.S. in the late '90s. Red is an attention-getter in some product categories, but can be a "red flag" in others. A maker of children's toys was about to use red as a graphic element for a packaged product, but had enough caution to test it with not only the kids, but with their parents. Research into emotional responses uncovered a concern by parents that the toy might be "too aggressive" for the age group, based on the red color. The packaging ultimately went with a green shade that offered shelf impact, yet avoided the "aggressive" perception.

How deeply emotional connections run

The way Coca-Cola rattled its consumer's psyche with an eye-catching white shows just how deeply emotional connections can run. The company regularly runs holiday promotional packages—Santa Claus, snowflakes and other seasonal cues. In 2011 it opted for a white theme for regular Coke. The color choice answered Coke's concept for a bold, eye-catching campaign; it also supported the beverage company's work with the World Wildlife Fund to build awareness of global warming's threat to polar bears on the Arctic. A month later, Coke went back to its iconic red cans in the face of consumer push back.

Sensation transference. As Louis Cheskin's precepts might have predicted, some people thought the beverage actually tasted differently in the white can versus the red can. Others said they confused the promotional cans with silver cans for Diet Coke. Coke replaced the white cans with a traditional red can.

Impact of social media. Social media certainly added to pressure on Coke to abandon the white can. Generally, consumers on social media liked the white can, but there was enough "noise" from those who didn't. Five weeks after its launch, the white can disappeared from Coke filling lines.

Key insights to consumer actions

Know that two factors drive the new consumer

One is the frugality imparted by the recession; the other is the wealth of information from electronic devices. But, don't forget that they hold onto basic needs that can be a tiebreaker in a buying decision or trump price when the need is met by a package.

Packaging Impact: Packaging design carries an even greater burden in communicating benefits and adding convenience, freshness and other attributes. All while hitting a price point.

The person who buys and uses your product reacts to packaging stimuli in different ways at different times

You have to know what moment you are trying to influence—from Zero to Third Moments of Truth.

Packaging Impact: A better understand of these moments can lead to better packaging briefs that encompass all the factors that influence development.

Build an experiential connection with the consumer

Packaging Impact: The emotional connection builds sales and retains brand loyalists. The package is a key vehicle to build that connection.

Use neurological tests to add deeper consumer insights

At its current level of sophistication, use neurological research to reinforce other insights rather than as a stand-alone research tool.

Packaging Impact: Design can get better as package developers measure the complex consumer decisions.

Know that consumers link package and product through sensation transference

It can be a basic sense such as flavor. It can be the perception that "this is the upscale product for me." Whatever the attribute, the package can influence it.

Packaging Impact: At the point of purchase, the package may be more important than the product because it defines the product to the consumer.

3 || *Opportunity in the Perfect Retail Storm*

> ***A scenario . . . Seoul, Korea.*** In a subway station. Ye-Jun Park peruses a wall with photographs of groceries, many of them packaged. With his smart phone, Park scans several 2-D bar codes on the photographs and puts them into the phone's "shopping cart." He presses "send," and his order is off to a retailer. After he gets home, a delivery drops off a carton containing his purchase. For Park, that's one onerous shopping task done. For international and private brand owners, the transaction adds complexity to their packaging decisions. The design function adds this criteria: "How well does the package photograph?" Sales need to assure good placement on store planograms and on the subway station display. And the packaging has to be efficient in distribution from retailer to home.

Retailers face a "perfect storm," and packaging is a key survival tool, both for retailers and brand owners whose products are on shelves. The storm is powered by lingering effects from the 2007 recession, continued economic uncertainty, and by changing shopper values multiplied by e-everything connectivity. Couple those "waves" with the relentless pressure on costs, and retail is being shaken up in ways it hasn't seen in decades. Walmart's same-store sales sag. Traditional grocery supermarkets lost 1.1% share of market in the U.S in 2010. There's a similar pattern in Great Britain where traditional grocers see sales lagging behind price inflation. And for every Walmart that opens, eight new "dollar stores" open their doors.

Here are drivers that are "bending" retail channels and how they affect packaging strategies for both retailers and brand owners:

- Retailers push to make shopping an experience, and packaging has to support that, both with graphics and innovation that pushes shelf impact.

- Retail channels are fragmenting and customizing to meet changing buying patterns. Packaging has to adapt to the new patterns.
- Retailers are moving into more locations—being where people want to buy. That includes being on line, in kiosks and in Korean subway stations. Packaging has new communication challenges.
- The PTIS Packaging Well Curve concept shapes a new direction for retail packaging. Real winners emerge from the "tails" of the curve.
- Private brands are paramount for retailers. Packaging development and innovation rises in importance among retailers. Brand owners facing private brand competition face challenges as the private brands move upscale and alter consumers' value perceptions.
- Retailers and CPGs need to gain even more supply chain efficiencies. Collaboration between retailers and brand owners has to support joint packaging decisions.

These drivers are global, and the packaging strategies to meet them are global, too. Just to put retailing's global retailing in perspective, here's some data from the Deloitte report, "Switching Channels: Global Powers of Retailing 2012." It lists the top 250 global retailers in 2010, with an analysis of revenue, retail sales and net margin. U.S.-based Walmart heads the list with US$419 billion in retail revenues. At No. 250 is Japan's Izumi Co., Ltd. with US$3.3 billion.

As we look at the data and how marketing and packaging strategies play out in the store, a number pops out for us. It is the margin figure. On a qualitative level, the margins suggest some positive correlation between strategic innovation programs and bottom line results. We say that because four of the top 10—Walmart, Tesco, Costco, and Walgreen—are firms that we categorize as strategic packaging innovators; they lead in using Big "P" Packaging as an integral part of their vision. Walgreen's probably wouldn't have been on our list a year ago, but its move to a cohesive strategy for its private brands puts it on the list. Its Nice! brand rolled out in 2011 with an emphasis on a cohesive positioning compa-

Organization	Home	Countries	Retail Sales, 2010 US$	Margin, 2010
Walmart	US	16	$419.0	4.0%
Carrefour	France	33	$119.6	0.6%
Tesco	UK	13	$92.2	4.4%
Metro AG	Germany	33	$88.9	1.4%
The Kroger Co.	US	1	$82.2	1.4%
Schwarz Unternehmens Treuhand KG	Germany	26	$79.1	n/a
Costco Wholesale Corp	US	9	$76.3	1.7%
The Home Depot	US	5	$68.0	4.9%
Walgreen Co.	US	2	$67.4	3.1%
Aldi Einkauf GmbH & Co.	Germany	18	$67.1	n/a

Source: Deloitte: Switching Channels Global Powers of Retailing 2012.

Figure 3.1. Top ten global retailers.

rable to national brands. Whether it is Walgreen's or any of the other leaders, we can't make a quantitative assessment because innovation and packaging aren't the only factors that determine margin—market category, region, expansion strategies and other factors influence margins. However, innovation and packaging are powerful levers in achieving better margins.

ADDRESSING THE DRIVERS

The kinds of margins we see from innovative retailers come from keeping a close eye on drivers that change retail. Here are details on those drivers and steps packagers and retailers take to gain from them.

E-everything shoppers still want an experience

The electronically enabled shopper we explored in Chapter 2 tops the list of drivers. Armed with computers and tablets at home and smart phones on the go, they impact packaging in three ways. One is the "let's have some fun" approach to shopping that demands packaging offer an experience. Another is the prolifera-

	2010	2017
Convenience	Store nearby	Click and buy
Service	Helpful	What I want, when I want it
Value	Good product, fair price	Best product at the lowest price

Figure 3.2. How consumers redefine their retail interaction.

tion of quick-read codes on packages; they help consumers know more about the brand, product and company. Finally, and it's a theme that runs through this book, is the cost pressure exerted by their hunt for value.

Jim Kauffman of the Everest Group is a marketing consultant we've worked with. He likes to say that the digital technologies are "game changers" at retail. He predicts shifts in the way consumers define convenience, service and value, summed up in Figure 3.2. Look, in particular, for signs of the "its all about me" attitude; it is a thought process that says, "I want what I want when I want it. I'm served, not targeted."

Generation M is driving changes and shaping the future for both retailers and packagers. Kauffman points to these trends in the store:

- Extended product selections. From a packaging perspective it makes brand differentiation even more important. And, as varieties grow, a design architecture that clearly differentiates varieties grows in importance. That applies not only to consumer packaged goods, but to foodservice as well. Don't forget packaging's role in highlighting the new offerings.
- Access to products through in-store kiosks and other points or through the Internet from home or mobile devices. Package design criteria—both graphic and structural—shift. A key question becomes, "How well does it display in a photo or on-line depiction?"
- Promotional communications as retailers gain data on their shoppers. The question for designers becomes one of addressing the niches consumer data identifies.
- Customer engagement—they will began to "follow" retailers they identify with.
- Convenience in terms of anywhere and anytime. Packaging needs to respond to changes in distribution patterns.

The "get smart, have fun and save" syndrome

The hunt for price—in terms of dollars, Euros, yuans, or other currency—hasn't diminished the consumer's pursuit of other values. Consumers still want to be delighted. The American psyche remains concerned with economic instability, yet frugality is tempered by other values—convenience, wellness, simplicity, affordable luxury and the environment. Meld the new trend of value with older habits, and you get a pattern that says, "get smart, have fun, and save."

From a packaging strategy perspective, that has played well for several retailers who bring value and still emphasize the fun part of shopping. Same-store sales figures reflect that. Target is up about 2 percent in 2010 in same-store sales. Costco is up about 5 percent. Walmart, which hones in on price, saw same-store sales erode about 2 percent. How do Target and Costco earn gains? We believe part of it is each retailer's strategy of letting shoppers "have fun and save." Target's marketing strategy is "cheap chic." The retailer uses high-style brands and design to add fun to value. The color of packaging for its Michael Graves home decoration line isn't Tiffany Blue. But it's not far off. Costco leverages a "treasure hunt" strategy, getting shoppers to look through the store for the next value that delights them. Packaging development needs to keep a goal of delighting the consumer and shopper, even though cost looms as a dominant factor.

Quick response codes

The e-enabled shopper is a target for quick response codes. These codes reach out for the shopper in the store just as they reach for the consumer in the home. In early 2012, packages were the second most scanned objects by the smart electronics, just behind magazines and publications.

Codes address the need for more information and the need to create an experience. Codes on packages can help make shopping an experience and help build an emotional connection. Think of a wine bottle with a coded hangtag that directs the smart phone to a site that "romances" the brand. It helps make shopping an experience. The codes first started to appear on packaging for high-end products, mostly because the smart phone technology is landing in affluent households first. Some research says a third of the electronic users come from households with annual incomes of more than $100,000. That income average is going to come down quickly, say observers in the consumer electronics industry. As prices of sophisticated cell phones drop, affordable units will appeal to consumers further down the economic ladder.

The bigger question is: Will the codes become a mainstream marketing tool, or a fad that disappears? We think the answer depends on whether marketers and packagers take a Big "P," strategic approach to using the codes. Implementing them is more than just putting them on the package. It is asking what communications goal we want to accomplish. With that goal in mind, then the package and the supporting websites need to address that goal. Packaging and web design answers could be different for goals of emotional involvement versus product use explanations. Advertising and other marketing communications tactics need

Leveraging on-pack codes

On-package quick-scan codes help marketers do things they could only do before with heavy advertising. Consider this case story on how SC Johnson leveraged electronics to market its Scrubbing Bubbles Automatic Shower Cleaner. The marketing message is one of convenience, but the details are not easy to convey: Consumers need to hang the dispenser on the showerhead and press a button after the last shower of the day. A pump powers streams of cleaning solution around the shower and onto the shower curtain. The consumer rinses down the surfaces in the morning and has a clean shower. This solution cost more than the typical pump bottle, and SC Johnson needed a way to explain how the product works and show its value. And the company wanted to do that at the point of purchase.

The solution was to add a 2-D barcode to the package. SC Johnson opted for the QR Code, one of the most frequently used of the codes. It links a web-enabled, mobile device to an Internet demo video that shows how the product works. The package also included simple how-to on using the code and explained how to get the app to read it. The package expands its role from being a silent salesperson to being a more vocal agent. It adds an emotional benefit and an in-depth convenience factor.

to support the objectives. If that produces an "AHA!" from the consumer, then the codes offer a benefit.

Fragmented, customized retail channels

Shopper expectations and a willingness to go to different places to meet those expectations drives one of the biggest trends in retailing—fragmentation and development of new custom channels. Retailer strategies are to move into niches and away from the herd in the middle. This thinking answers the dichotomy in consumer thinking—sometimes I want price, sometimes I want something else—convenience, nutrition and health, luxury or another experience.

What retailers are doing with success is finding those niches that appeal to specific consumer buying needs. It's no surprise that the rise in the search for value and the increasing number of quick trips parallel the growth in dollar stores. It is just one facet of increased channel fragmentation that gets more intense as retailers better understand the consumer. For packaging, that trend means that a product and packaging strategy has to embrace retail fragmentation as a given.

Here are some of the ways those drivers manifest themselves. First, mass discounters and price-sensitive retailers such as ALDI and dollar stores aren't the only places they go. Because the value equation is not just low price, many mainstream shoppers indulge in small luxuries—chocolates or coffee from upscale retailers. And remember affluent shoppers—they rebounded from the recession quickly. Consultants who follow retail sales estimated that in 2009, sales of luxury goods dropped 9%. However, in 2010, estimates were that sales of luxury goods rose at least 10%, with some marketers of these goods seeing sales increases into the mid-teen ranges. The balance between price and other values plays out within sections. Consumers reduce spending in certain aisles but maintain or even increase it in others. Research shows that post-recession shoppers are more thoughtful about the brands they purchase and the type of products they need.

Even when price is the driver, quality remains very significant for American shoppers. According to an IBM study, 72% of respondents are more concerned with the quality of the food they're buying than the price. Additionally, nine out of 10 say that value as well as nutrition will be of equal or greater importance after the recession. Convenience, tamper-evidence, eco-friendliness are other drivers that consumers value.

Where major channels are going

Shifts in consumer buying habits form a background not only in the U.S., but in other developed markets. In Europe, retailers such as Tesco, Marks & Spencer, Carrefour, and Metro are also testing and implementing new formats. These changes extend from the home countries to retailing operations around the world. Here are major channels undergoing significant changes:

- Discount, mass merchandisers
- Club Stores
- Dollar Stores
- Traditional grocers
- Traditional drug stores
- Category killers
- Foodservice
- Internet enabled retailing

Discount, mass merchandisers. In the U.S., think Walmart and Target as the most visible retailers in a segment driven by price and high volume. The hallmark tactic within the segment is aggressive buying techniques. For packaging, that means pressure on costs. Yet, packaging has to respond to other demands, in particular each retailer's efforts at cost control within their own distribution and stocking operations. They know that each time an employee touches a package, it adds to costs. Walmart once calculated a cost per "touch" and is one of the drivers of retail ready packaging, which we will go into depth later.

Discount and mass merchandisers have strengthened their commitments to private brands as a way of building exclusive relationships with shoppers and driving higher margins with lower overhead products. The impact on packaging includes a greater role for retailer-controlled packaging design and innovation as tactics to gain exclusivity through private brands.

Club stores. Three major players dominate the club store business in the United States—Costco, Sam's Club and BJs Wholesale Club. Costco and Sam's are multinational in their reach. They charge a membership fee, limit variety and sell products in larger units. The impact on packaging is the use of display pallets and multipacks.

Club store economies and market position means retailing is from pallets on the store floor. They sell a limited variety, focusing on the highest volume SKUs and turning SKUs on a seasonal and promotional basis. A club store may have 6,000 SKUs on the floor at any given time, compared to the 25,000 SKUs of the traditional grocery store. Virtually no "touches" of individual packages by store employees. In the best-case scenario, a lift truck drops a full pallet on the floor, and the employee removes the empty pallet when it is sold out.

Yet, these stores want promotion and excitement, and that means packagers have to build promotion and pizzazz into the pallet itself. In a decade-plus, that need has stoked growth in the packaging value chain with the increased reliance on contract packagers. They take a typical grocery store package and put it in a multi-pack and array the multi-packs onto display pallets that drive sales velocity through the club store. This packaging strategy changes from a least-cost pallet approach to one where brand owners often have to invest in printed shippers, sometimes with novel display characteristics. The balance requires packagers to develop pallet stacking patterns that make the most efficient use of space while displaying packages well.

Dollar Stores. It's a simple formula—price plus convenience. Dollar stores appeal to the immediate purchase needs of people with a limited number of dollars to spend. Just look at the wall of laundry detergents common in dollar stores. This channel put special demands on packaging in terms of size and sometimes specific package graphics for the retailer. At times, major national brands such as Tide have had a special dollar store size; the package size is calculated based on the price tier the dollar store chain uses and then works back to a size that meets that price point.

In the first decade of the 21st Century, dollar stores were the hottest growing retail sector. They grew when the economy grew, and then they grew when the economy soured. One threat to profitability: as they add grocery items to drive traffic, they reduce their profit margins because grocery is a low-margin category.

Traditional grocers. They offer a wide assortment—essentially saying, "You, the shopper, can get everything you want in a single trip." Typically, a store displays 25,000 stock-keeping units (SKUs) or more. Here's a specific example of how that plays out for packaging—a grocer carries a dozen brands of pasta sauce, and carries several varieties and sizes in each brand. That's a hundred or so SKUs in one category alone.

That drives packaging design. The traditional response has been with graphics, but today it is likely to include structural packaging elements. The goal is shelf impact and differentiation from other brands in the shelf set. In addition, design architecture has to help the shopper's choice among varieties in a brand's line. The question is: "How can I help the customer quickly see the different varieties in my line so she doesn't get home to find she's grabbed the mushroom and basil variety instead of the traditional flavor?"

The commitment to wide choice, slow-moving SKUs and multiple levels of private brand products has pushed the traditional grocer into the middle of the PTIS Packaging Well Curve. New retail formats at the "tails" of the curve siphon off some of the retailers' highest-margin lines.

Traditional drug stores. Drug chains are expanding product lines to include more grocery and fresh products. Think in terms of drug stores positioning themselves as convenience stores for women: about 75% of their shoppers are women. In general, traditional convenience stores aren't friendly to female shoppers, and drug stores look to gain from that scenario. Packaging's role is to appeal to women with attractive packaging, even for fresh produce items. Drug stores are also moving into private brands to differentiate themselves.

Category killers. They give shoppers everything they could ever want in a product category. Think in terms of Staples for office supplies and PetSmart for pet products. From a packaging standpoint, the challenge is to "go deep" in the sizes and varieties offered. While a discount-mass merchandiser may carry a couple of varieties of pet shampoo, a category killer may carry three tiers of pricing—premium, every-day and value. And it may do that for several brands. Packaging size, structure and graphics are all factors that shape a packaging response.

Foodservice. Foodservice packaging is another area where

demands are rapidly changing. These changes are fueled by many of the same pressures that are applied to consumer packaged goods. With food away from home accounting for 35.4% of all food expenditures in 2010, it is a significant portion of consumer's expenditures. However, the category has been in decline, down from its high of 37.3% in 2006. The economic decline that began late in the last decade continues to influence consumers to trade-down, altering where and how much money they spend.

Overall, the restaurant industry had sluggish growth of 1.8% according to the Technomic 2011 Top 500 Report. The same report indicated a slightly better picture for quick-service restaurants (+2.2%) and fast-casual (+5.7%). Sales were essentially flat for full-service chains. While growth has been slow, the industry showed positive growth, something it failed to do in 2009, where sales fell by 0.8%.

We spoke with Shane Bertsch, Vice President of Global Packaging, HAVI Global Solutions. Shane and his team have extensive packaging and sustainability experience. They have identified several macro-trends that are impacting packaging in the foodservice industry:

- The Hunt for Value
- Consumers Push for Convenience
- Mobile Engagement
- The Sustainable Edge

Many of these macro-trends also influence consumer packaged goods retailers. Consumers hunting for value aren't anything new. Consumers are pinching pocketbooks, and finding the right combination of size and price has been an important component of foodservice strategy. Packaging is being stretched to cover a broader range of offerings, ranging from Costco-esque servings of sharable entrees to tiny value-sized snacks. This increasing range of offerings has added complexity to store operations as well as packaging sets.

Increasing complexity in consumers' lives is driving the need for more convenience in eating establishments. On-the-go eating is a major part of menu development; effective packaging must accommodate this trend. Nowhere is this more evident than in quick-service restaurants, where drive-thru and take-away orders continue to dominate the segment. Similarly, consumers' eating habits continue to shift outlets such as carts, kiosks, and food trucks serve in-between snack occasions that don't line up with traditional meal times. Packaging for these limited service locations that still provide differentiation is a significant need.

Consumers have never had more access to information. Packaging can further effectively engage this trend by interfacing with mobile devices. Providing health information, coupons, or cross-promotions are just a few examples of ways to integrate packaging into a business's overall marketing strategy. Finally, though certainly not least, both foodservice and consumer packaged goods companies must have a plan for sustainability. This is especially apparent in the quick-service segment where waste and perceived over-packaging are highly visible issues. Putting emphasis on reducing the amount of packaging used and developing alternative end-of-life scenarios to landfill are crucial to success.

Internet-enabled retailing. We could simply say, "Internet" retailing, but we're already seeing newer forms of electronically enabled retailing that are redefining the category. Just how "hot" is this category? Interbrand's 2012 Best Retail Brands Report puts Amazon at No. 9 on its list with a brand value of $12.8 billion, just behind Sam's Club with a brand value of $12.9 billion. Amazon led all retail brands on the list with a 32% increase in value from 2011. Moving into the top 10 is eBay with a brand value of $9.8 billion and a 16% growth rate. eBay's rise puts two e-tailers in the top 10 brands on the list.

Both of the Internet-based organizations have honed in on packaging as a branding component. Amazon.com has a comprehensive program in asking for Internet-specific packaging. Like savvy bricks-and-mortar retailers, Amazon looks at what its shoppers want and what its retailing format needs to be successful. Several years ago it rationalized its own distribution packaging; it devised effective, simplified shippers that addressed consumer complaints about excessive packaging. More recently, it is addressing complaints it hears on clamshells as an element of what consumers consider "over packaging." Clamshells make a great deal of sense for bricks-and-mortar retailers where the packages drive impulse buying and deter theft. In some cases, theft runs as much as two percent of sales going out the front door with shoplifters. Clamshells make it difficult for shoplifters to remove items from the package and carry them out of the store.

But Amazon is a different retail operation. They don't have customers shoplifting, and they don't need packaging to drive impulse buys. Clamshells do little for Amazon, except irritate its consumers. So it asked vendors for alternative packages and got cooperation from makers of brands such as Duracell batteries and Fisher-Price toys. Nothing fancy, just replacing the blister packs with efficiently designed corrugated paperboard shippers. What it means for packaging is creating another SKU to meet the needs of a specific channel and the development costs to create the new packaging.

Amazon's tactics underscore the relationship between pack-

Eco-friendly eBay Box needs strategic support

EBay's eco-friendly box shows that for Internet, too, packaging works best when fully integrated into a strategic business plan. In summer 2010, eBay distributed 100,000 corrugated shippers—in four different sizes. The goal was to have a reusable box that eBay sellers and buyers could use over and over to reduce resources consumed in shipping. Graphics—using water-based inks—touted the better environmental footprint and encouraged reuse. The box structure even required less tape to seal. Yet, by summer 2010, when eBay ended the test, it tracked less than 1,800 shipper reuses.

"We know anecdotally that far more than 1,800 of the boxes were reused but learned very early that asking people to manually track their boxes on a website wasn't going to be the right tracking mechanism," said an eBay rep. The e-tailer's tracking system had people input data onto a website. EBay says it remains committed to the strategy but needs to support the physical packaging with better IT and ways to promote environmentally friendly packaging.

The belief has some validity if you look at the FedEx reusable legal-size envelope. Its structure lets it be refolded and used a second time, and FedEx targeted it to certain industries such as publishing. With FedEx's tracking sophistication, the envelope has been in use since 2003.

aging and branding from the retailer's perspective. Amazon, like its brick-and-mortar cousins, is a brand. And the packages it now gets to replace clamshells supports the brand. They are neat, functional with no-frills to reinforce Amazon's position as eco-conscious retailer. Amazon has even trademarked the phrase Frustration-Free packaging and offers a Frustration-Free certification for packaging that meets its criteria.

Amazon's strategic use of packaging to reinforce its brand image extends to the independent sellers that use amazon.com as a channel. Amazon, and competitor eBay, gives sellers packaging guidelines that stress professional, clean packages. The web retailer goes so far as to recommend metered postage and discourage the use of small denomination stamps, practices that raise the credibility of the independent seller and Amazon.

The "frustration-free" bandwagon is gaining momentum with actions by Walmart and Tesco. The UK-based retailer has gone after the wire ties used to position toys within cartons.

We believe the growth of electronically enabled retailing will have an impact on package graphics. Much of packaging graphics respond to the First Moment of Truth we detailed in Chapter 2. It is the instant the shopper meets the package at the store shelf. But, what about the shopper who meets the package on-line? Often, design intricacies that signal brand positioning and attributes simply don't have the same visual impact on the screen. It's the same question that faced catalog marketers for decades and led to this criteria for catalog brands: "How well does the package photograph?" We anticipate that to happen with Internet-specific packaging.

Other channels

Both developed and developing countries are seeing small format outlets grow—those include kiosks and bodegas (or corner stores). In developed economies it delivers geographical convenience; in developing economies it may be the first step in developing a retail infrastructure. The impact on packaging is the need to protect brand equity, anti-theft and anti-counterfeiting threats, and product protection in environments that require more durability. These retail formats also ask for smaller sizes to fit the purchasing ability of consumers who don't have a lot of money.

These channels also push packaging's protection function; with smaller sizes and extended distribution, the need to preserve freshness and prevent spoilage are more acute.

Vending offers intriguing growth potential. Today, coffee machines dominate sales volume, but global innovation has stretched the potential to include made-to-order, ready-to-eat pizza and iPods. Even for conventional packaged products, vending impacts packaging: Will a beverage container fit within vending machines? That dimension may determine the dimension on on-the-shelf packages. In some cases, vending may even require manufacturing and packaging within the machine, adding package development and quality control requirements.

Virtual shopping in a Korean subway station

The "gee-whiz" angle of virtual grocery shopping has piqued the imagination of retailers. One approach, described at the beginning of this chapter, is to allow shoppers to scan codes on virtual packages arrayed on a backlit billboard. The example is in a subway station, but it could be anywhere that generates enough traffic. Shoppers capture codes on their smart phones, relay them to the retailer, and get their groceries delivered to them. This retail innovation goes well beyond the "gee-whiz" factor. It is based on research, business objectives, and technology to build an innovation platform that served a business need and met a consumer need. The retailer is Tesco, operating as Home Plus in Korea, and here are elements of the platform.

Business objective: Despite having fewer stores than its competitor, could Home Plus become the No. 1 retailer in Korea without adding stores? Could on-line shopping do that?

Consumer research: Koreans are the second hardest working consumers in the world. Grocery shopping once a week is a dreaded task.

Technology: Virtual displays mimicked store shelves. Smart phones provided the interface.

Results: Tesco says it had 10,287 people visit its on line store during the promotion. It got a 76 percent rise in the number of people registered on its site. And on line sales rose 130 percent to become No.1 in Korea. A "halo" effect increased sales in bricks-and-mortar stores, too.

Retailers within channels

In some of the channels, the biggest players demand their own packaging. Look at club-store competitors Sam's Club and Costco. In that competition, Costco wants to sell larger quantities, so it carries Kraft Macaroni & Cheese dinners in a 15-pack. Sam's Club's multi-pack is a 12-box unit. Packagers need to bundle, palletize and ship based on the specifications of a particular retailer.

Target "gets it"

If we have to cite one retailer that "gets it" it is Target with its shopper-centric focus for both its retail environment and for packaging that supports the environment. Consider household décor items like the Michael Graves line. Products demonstrate design that meets the Target "cheap chic" aura. And the blue boxes that hold Michael Graves' products aren't that far off in color from Tiffany blue.

Target is about design, and that gives it its cachet in the discount-mass channel. Consider the Method brand of household cleaners launched through Target. Method's differentiator is its role of bringing design into the home. Products and packaging like that are perfect fits for Target's retail platform. Target knows the power of packaging to sell products. It has its own packaging development staff, and applies that strategy to its private brands. It is an example of the retailer drive to exclusivity through private label packaging. Target has multiple private brands, to appeal to different shoppers within its stores. The Market Pantry brand for foods is close to the traditional store brand that offers every-day savings. Target's Archer Farms brand is at the high-end of the PTIS Packaging Well Curve, offering exclusivity to the retailer.

One of the classic case stories of retailer brand differentiation is the Clear Rx prescription drug packaging system. It blends Target's overall strategy of communicating with consumers and its strategy of using design to project value. On the communication side, the bottle provides an information hierarchy aimed at more effective communication with the consumer. The drug name is clearly printed at the top, with dose instructions below that. Administrative data—quantity, refills, doctor's name—is

	Target	Walmart
Gross Margin	31.64%	24.96%
Operating Margin	7.80%	5.95%
Net Margin	4.31%	3.54%

Source: Seeking Alpha.

Figure 3.3. Margins of Target and Walmart.

in smaller type below that. The container itself is red, Target's brand color that is also on the cap. The package uses color rings above the closure so different family members can quickly see which bottle belongs to which family member. In retail, where marketers want distinguishing products that support their market position, this is an exclusive for Target.

This case story is also a great example of open innovation. The concept was the brainchild of Deborah Adler, a graphic designer who developed the Clear Rx concept for her master's thesis. In the simplest terms, Target saw it, liked it, bought the patent and pushed it to commercialization as fast as it could. We'll give you ways to build open innovation into your business strategy in Chapter 9.

A final note of how Target's marketing and packaging strategy pay off for its business. We know that margins are just one metric in analyzing a business, but Target has higher gross, operating and net margins than Walmart. In a review of four recent quarters of results, all of Target's margins are higher than Walmart's. It's just one metric, and a lot of market analysts will say Walmart is still the preferred investment. But, say others, "It's a lot closer than you think."

The PTIS Packaging Well Curve

Amid this divergent backdrop, we see a model that points toward profitable options in both retail and packaging. The model is called the "well curve." We found the basic concept on the website of Jim Pinto, a California business consultant. If you haven't heard this term, it takes the bell curve concept and turns it upside down, both literally and from a conceptual view. Where the bell curve suggests that preponderance of the action is in the

Figure 3.4. Grocery retailers and PTIS Packaging Well Curve.

middle, the well curve suggests that profitable action is at the ends of the curve.

Let's look at how that translates into the retail scenario. Here are three formats in retail foods: Walmart, Jewel-Osco (the Supervalu brand in the Midwest U.S.) and Whole Foods. Walmart is positioned on the curve's left side and thrives on a deep discount format; it focuses on price and has seen growth and profitability in that niche. In the middle is Jewel-Osco; in its midrange niche, it is in the Well Curve's danger position. It has seen its market share decline. On the Well Curve's right side is Whole Foods. The Texas-based retailer focuses on organic and natural foods and operates in the U.S., Canada and the U.K. It thrives on

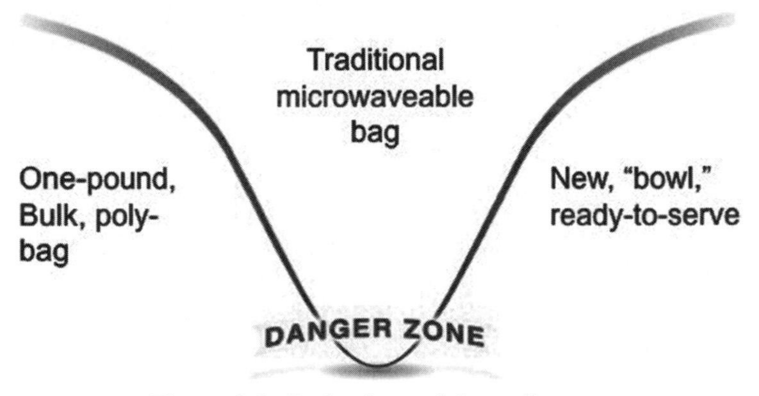

Figure 3.5. Packaging and the well curve.

the premium side of the curve. Consider Starbucks and Target as other retailers embracing the premium side of the well curve. Growth and profitability align with position on the well curve. These retailers use packaging to support their position on the curve and look for innovative thinking in that packaging.

That same thinking applies to packaged products. Here's just one scenario to demonstrate: packaged popcorn. On the left-hand side is bagged popcorn. It is basically a bulk purchase with the minimum of packaging—a printed poly bag. It is the frugality positioning. In the middle is the microwave popcorn bag. At one time it was an innovative solution, but age has pushed it to the center of the curve where it is undifferentiated and squeezed on profits by the number of competitors and SKU proliferation to accommodate more and more flavors. On the right-hand side—in the premium position—are the new "serving-bowl" packages. When they pop, the package forms into a serving container that can sit on a table so people can nibble from it. It is today's innovation that brings added convenience consumers want.

Exclusivity drives private brands

Retailers increasingly look to differentiate and stand out from the crowd. One of the strongest levers they have in doing that is "private brands." You may have heard the terms "private label" or "store brands," but we think "private brands" is the better term. That term says there is sophisticated branding and packaging that savvy marketers use to differentiate themselves, not only from other retailers but from national brands. Private brands give shoppers a reason to keep coming back to a specific retailer because that is the only place to get the brands.

Europe sets the pace for private brand leadership with their importance varying around the world. Data from Nielsen shows Switzerland leading in private brand market share with about 46% in 2010. The U.K. is about 43%. In North America, Canada has about a 24% share and the U.S. is about a 17% share. In the U.S., private brands saw about a two percent rise in market share since the Great Recession began in 2007, and many say it's the poor economy that is driving sales. Yes, that is one factor, but if you look at retailer strategies and what's going into private brands, there are other, deeper drivers for the growth.

And those drivers often find an ally in packaging to make them come true.

In the U.S., private brands are growing faster than national brands (1.7% versus 1%, according to Nielsen) and they represented $88.5 billion in sales in 2010. Packaging is a big part of their success and will play an even more important role in the future as they incorporate the convenience and visual appeal formerly in the realm of national brands. Today you see new package formats, including stand-up pouches with zippers, shrink labels on bottles and high-end canisters for dry products. Private brands benefit from the explosion in printing capabilities. What used to take lithography to produce can now be done less expensively on a flexo press with a comparable quality. That's just one example of better quality across the entire spectrum of printing.

Packaging is adding to the overall product value equation through enhanced graphics and benefits (for example, enhanced freshness, convenience or microwaveability). That points to additional growth opportunities across product categories and retail channels. We have heard from credible sources that private

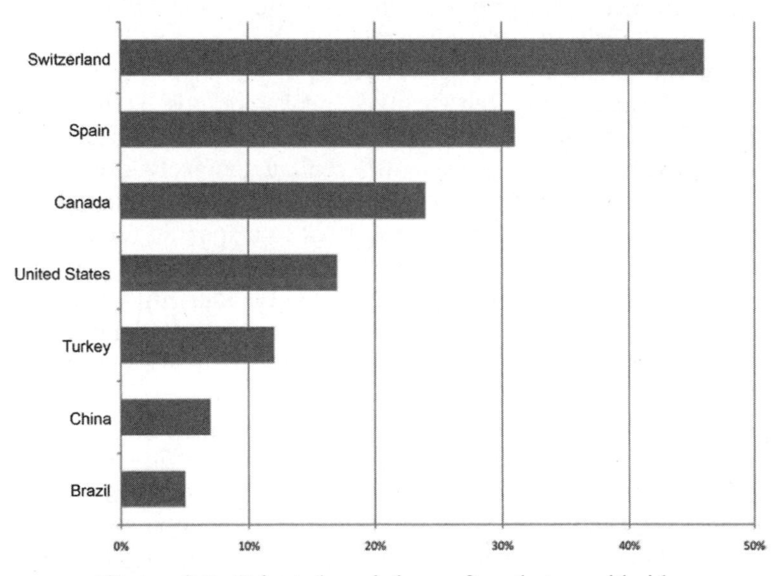

Figure 3.6. Private brand share of market, worldwide.

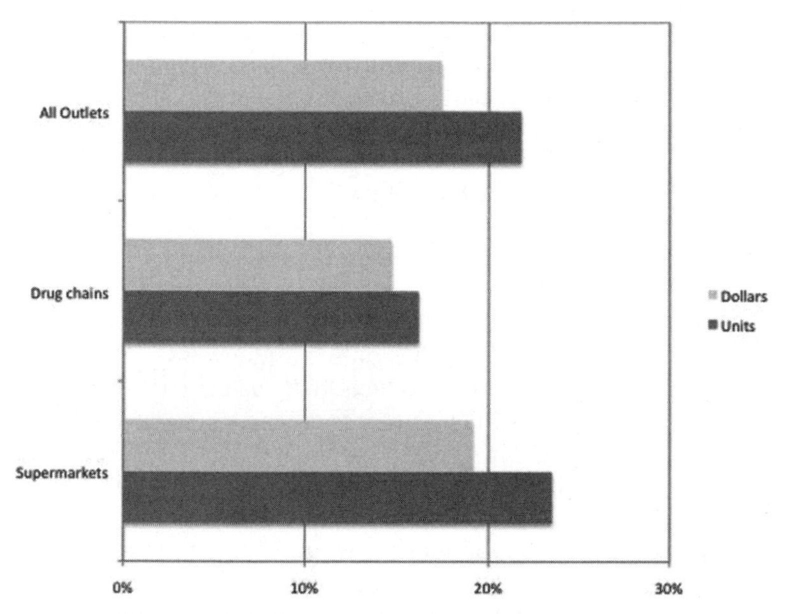

Figure 3.7. US private brand share, by outlet type.

brand growth in the United States could double in the next 10 years; packaging, with its ability to communicate market position and value, is a key enabler for this growth.

Private brand packaging checklist

Here's a quick checklist that private brand developers can use to assure that packaging makes the maximum contribution:

- Graphics should support brand image and essence.
- The packaging's features/benefits are equal to or better than the national brand.
- Packaging is on-trend for consumers and customers, as evidenced by convenience, freshness, portability and retail readiness.
- The packaging meets universal design considerations for ease of opening, usage, readability and other points.
- The cost-to-value parameters have been developed and are understood internally.

- The environment/sustainable packaging story and end-of-life scenarios have been developed in terms of recyclable materials, design for disassembly, use of renewable materials and adherence to FTC Green Guides.

Retailers who do private brands well

Trader Joe's positions itself as an upscale retailer of quality, innovative products; it also emphasizes customer service and a "neighborhood" feel as a differentiator. Trader Joe's carries no national brands. Rather, it relies on the uniqueness in its private brand and a succession of products that makes each trip to its stores an adventure. Aldi is a sister company to Trader Joe's, and relies on private labels to achieve and brand itself as a price marketer. Its private brand strategy is very much like the European experience.

We've talked about Target, and we need to emphasize its Archer Farms brand. It is a high-end brand that offers quality and supports the Target brand positioning as a retailer that brings value through design. Consider Archer Farms organic cereals and its packaging. Like any brand, the need was to differentiate it from competitors. The carton leverages both graphics and structure to set it apart.

As we've said throughout this book, packaging is not an entity unto itself—it is part of a business strategy. In developing that thought process, consider IKEA. Here's a company that neither "sells" nor highlights the value of its packaging, yet it is an integral part of its business strategy. *The key point is this: The packaging for its "knocked down" furniture line enables the business model.* It provides both efficient distribution from manufacturing points, and it is consumer-friendly in transporting the furniture from store to home. The packaging makes IKEA seamless as both a retail and product brand.

Supply chain efficiency: Retail-ready packaging

Packaging adds impact to the store shelf, yet it is also a "soldier" in the war on costs. The store shelf is one small battleground in that war, and it is a site of an action that reflects a larger strategy called "retail-ready packaging".

In simplest terms, it means reducing costs at retail by cutting the time it takes to restock a store shelf. Having a stock person cutting open cartons and putting individual packages on shelves isn't the way to do it. But having the stock person put an entire tray of 18 cans on the shelf in one motion is an answer. Making that happen also is part of a bigger picture that extends back to the stock room, forward to the shopper interface with the shipper and ever further into the future when the retailer disposes of the shipper. Success requires close collaboration between the packager and the retailer, and both need to look at it from a system approach that can yield overall savings.

Leadership is already coming from Europe where the retail-ready packaging concept originated; ECR UK is a consortium of businesses driving the European solution. It developed the prototype definition of retail-ready packaging. That definition extended to Canada where Walmart and Loblaws are in the process of implementing it. Here are five principles that define retail-ready packaging.

Clear identification of brand, product type and variants on the case for the store staff that is stocking shelves. That includes employees searching the back room for the right shipper with the right sized container in the flavor that needs to be replenished. The employee needs to do that quickly among a sea of similar containers.

Easy opening of the case to reduce the time needed to open shelf packaging. The process needs to be clean, produce a "best" finish, and reduce the need for a cutter. Especially as retail-ready cases go onto store shelves, a poorly cut edge can transfer a sloppy, unfinished image to the product inside.

Replenish with a 'one touch' movement onto the shelf. It is faster than handling each unit within the shipper. "One-touch" stocking has been around for a while, and the term PDQ (pretty darn quick) trays described it. However, PDQ's put little emphasis on graphics and shopper communication described in the next criteria.

Easy to shop calls for at-a-glance recognition of the category and unimpeded access to the product by the consumer. This involves both structural design and graphics on the shipping container. Structurally, how can the case protect the product in distribution, yet display primary package graphics on the shelf?

This is a new criteria arising from "lessons learned" in earlier "pretty darn quick" (PDQ) efforts. What PDQ's lacked was an emphasis on making it easy for consumers to shop a category and develop an emotional connection with brands. A few years back, the authors did a store audit of a retailer with a premium private brand product; the brand had sophisticated packaging graphics to convey a high-end look. However, the retailer also wants stocking efficiency and put its sophisticated primary packages in brown, undecorated PDQ trays that diminished the primary container's graphics. And the retail went one step further to detract from the image. It used a large format dot matrix printing on the lip of each tray to carry variety, date and lot information. What the consumer saw was conflict—one set of graphics said "premium," the other said "mundane."

Easy disposal of the distribution packaging. Cases or trays should be straightforward to dismantle, to separate components and to recycle or return materials. The idea of sustainability dwells here, too.

Trials in Canada have already encompassed the candy category. Walmart described its results in this category as "pretty good." Retail-ready packaging also appears to support the retail trend toward smaller stores. Unit sizes may be smaller, but the concept can bring efficiency to smaller stores that really need to emphasize economies.

Do a competitive retail audit

We've packed a lot of strategy and insights into this chapter, but here's a down-to-earth way for you to assess how your packaging—and your product—stacks up against the competition in the store. The tactic is the "retail competitive audit." It is a team process and it gives your team one of the best snapshots of where your packaging fits into the on-shelf "pecking" order. It can uncover strengths and weaknesses you might not see otherwise.

The audit is a management tool, and there are a few management caveats before you head to the store. One is to identify who is the steward of the data you gather. This process demands a repository for the information to keep the work's value from getting lost. We've got a point of view and think the packaging manager is a good place to put this function. Along with marketing people, the packaging staff belongs on the team to keep the packaging lab in the loop. Their input can be valuable, especially in analysis of different packaging technologies. Once you've got the team in place, use these steps to gain usable information:

- First, list all the retail distribution channels your product is sold through. The channels outlined in this chapter form a good checklist. Then identify the categories you need to audit. They are the categories in which your products compete. Finally, make a list of competitors you want to track.
- Record the audit, because you need a tool to define any actions on your part. Use a simple spreadsheet, or use a table in a word file. Key headings list competitors, and the relative advantage of each package, and identify significant advantages. It also includes recommended "next steps." With the ready availability of cameras on phones, you might want to incorporate photos from the shelf into the audit. The steward makes sure the data are updated.
- The steward circulates the data to each team member. Put the insights, recommendations and next steps on the agenda to get everyone's input into any revised directions.
- Get a bonus by auditing other categories. Find one that is different but shares some common packaging characteristics with your category. If you're in birdseed, for example, look at dog food. You may find some ideas that can work to give your brand an edge in your category.

Key action points for retail channels

Know how smarter, mobile shoppers leverage mobile electronics in their hunt for value

Know that they use those same devices to find products that satisfy their need for individuality.

Packaging Action: Innovation needs to deliver the "delight" factor shoppers want while still addressing cost concerns.

Focus on the First and Second Moments of Truth

The first gives you an edge on the store and shelf; the second brings consumers back after a satisfying use of the product.

Packaging Action: This involves both graphic and structural design. Innovation may be required to provide the differentiation. Packaging research needs to focus on the shopper experience. From a visual perspective, assess your packaging from a distance and against the category's context.

Keep PTIS Packaging Well Curve in mind

It is a model to gain profitability by identifying unique, niche offerings with business value.

Packaging Action: Design—both graphic and structure—has to communicate the product's positioning along the curve. When packaging adds convenience or "gee whiz" factor, the package itself may become the niche-defining element.

Follow how retailers are fragmenting

It is the strategy they are using in their struggle to be profitable with smarter, mobile consumers.

Packaging Action: Packaging and business strategies have to be flexible to accommodate changes. Practices such as contract packaging can be more effective in addressing changes.

Know that private brands are an effective weapon in the struggle to gain profits

Executed well, they can increase margins and offer a retailer exclusive brands that shoppers can't get elsewhere.

Packaging Action: Both retailers and brand owners need to look to packaging innovation to help define and communicate benefits of their brands.

4 || *Answers Are in the Value Web and Teamwork*

> ***A scenario . . . Dallas.*** The bachelor who tossed the frozen dinner into the microwave didn't particularly see any difference in the tray versus the one he threw into the microwave a week earlier. Yet, innovation by the food packager, tray supplier and raw material suppliers improved the tray's carbon footprint and cut its cost with a filler. As an innovation, it is mid-level on the scales of risk and impact, but underscores just how the integrated value web supports innovation. Let's look along the value web and see who drove this innovation. The impetus came from special interest groups and retailers who want to cut packaging's environmental impact. The food packager was the focal point and also engaged the value web. First, the company contacted its tray supplier; at the same time it went to a material supplier and an overseas university think tank. The global resources across several levels of packaging made the innovation happen.

Packaging's reach, as a business function, is far, wide and complex—so much so that we refer to its reach as the "packaging value web." It grew from a simpler value chain model in just more than a decade; that growth is a major reason packaging has become a strategic business function. Here's our premise: The process of packaging development demands collaboration within a company and outreach to external resources. In doing that, packaging becomes an enabler for other business functions. Consider these connections that packaging makes regularly that enable other business functions, both inside and outside an organization:

- It collaborates with groups that influence business plans. Packaging opens pathways to innovation, sustainability and cost-control.

- Packaging helps set strategic sourcing paths. The contract packaging option, for example, leads to cost analyses of production options and can be a "go-no-go" factor in deciding to launch a new product.
- It builds a pathway to technology and innovation. In opening routes to those areas, packaging helps to address the changing business needs.
- Through transport packaging needs, it joins the logistics network; it can lead to collaboration with third-party, and even fifth-party, logistics providers.
- It facilitates global options.

The value web addresses business uncertainty by giving you more knowledge for reasoned decisions. The links become an "insurance policy" to cope with uncertainty, and your goal is to know how every one of the "players" in the web can create risks or help you with answers that address your strategic needs. This value web serves both the packaging function, and it is also an enabler for other business functions. We think the best way to see how this web can help you is to take an in-depth look at it and some of the ramifications of working within it.

KNOW THE NEW INFLUENCERS

The value web, as an interrelated group of external organizations, has a great deal of influence on your product, packaging and brand character. You control that influence with your understanding of how the value web works. It has grown increasingly complex over the past decade. Let's start with how it worked in the late 1990s.

Five core groups shaped the packaging development process. It was linear and largely went in one direction: Raw material suppliers sold product and new ideas to converters who used them to make packages sold to brand owners. Retailers took what the

Figure 4.1. Packaging value chain in the late 1990s.

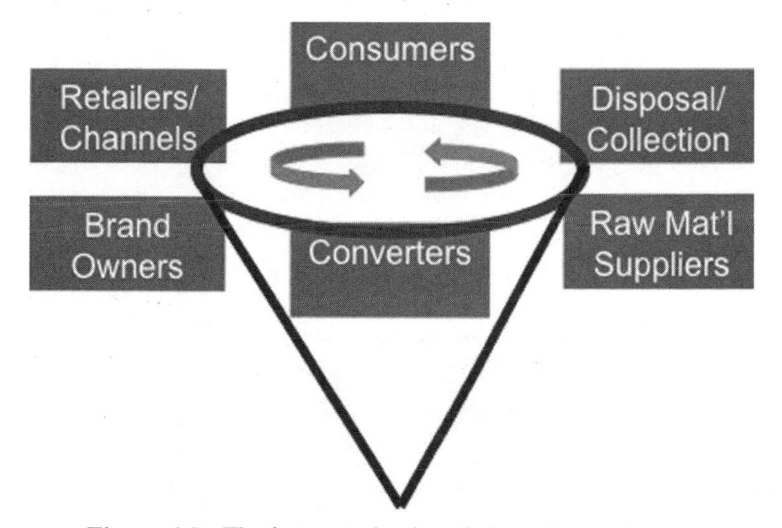

Figure 4.2. The integrated value chain and core players.

brand owners offered and sold the packaged products to consumers. Yes, there was some occasional collaboration back down the chain, and in rare cases, someone like a raw material supplier would go to a consumer to see how the packaged product really met the consumer's needs. But not often.

Growing consumer "clout" and a heightened emphasis on branding forced the value chain into a circle that was dubbed "the integrated value chain." Everyone began talking to everyone else and the knowledge flow was no longer linear. Converters saw the role of consumers in shaping their offerings. Notice, too, that the function of "disposal/collection" earned a place in the chain. The idea of eco-impact has been around for a while, but its rising importance finally pushed it into the process.

Then, the first layer of "influencers" began to have more impact. Among the changes: Packaging design sometimes led—rather than being the last group to have input on an innovation. Retailers, in a highly competitive situation of their own, began to use packaging as a strategy; in doing so, they expanded their influence. Sometimes it even had an impact on packaging technology. The Internet's impact on packaging grew, and a base for social media formed. Equipment and machinery had more influence, too, as advances in digital controls added more flexibility and efficiency. Non-government organizations (NGOs) work in tandem with government; they gained influence, particularly in the eco-friendly arena. The government's role is clear and growing with packaging bans and deposit laws on local, state and na-

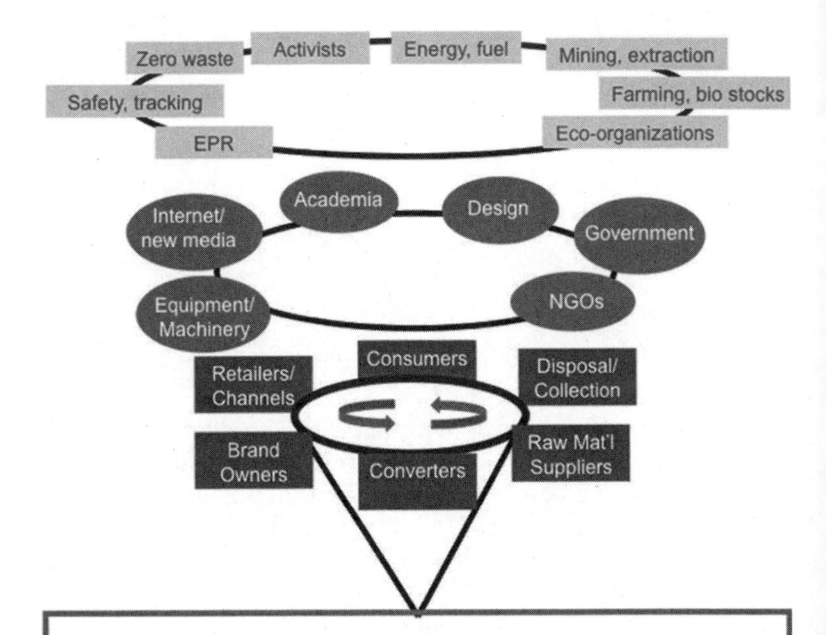

The value chain morphs into the value web

The term "value chain" has been part of packaging jargon for decades. Even as relationships became more complex, they often played out in a linear fashion. We're suggesting the term "value web" to describe what is really going on now. Relationships extend well beyond traditional paths: NGOs, sustainability and raw material suppliers have more impact on packaging development. We think calling that the "value web" helps us to better understand and manage the process.

Figure 4.3. Integrated value web.

tional levels around the world. Expect academia's influence to grow, particularly in technology development. For NGOs, look at groups such as the Toxics in Packaging Clearing House—it aims to reduce heavy metals in packaging. GreenBlue, and its affiliate the Sustainable Packaging Council, is an NGO with positive ties to the packaging community as we continue to evolve the framework for sustainability. As states cut funds for universities throughout the United States, the institutions are looking at private revenue sources. Packaging technology is a fertile area. Even with their financial struggles, universities find ways to hire researchers and build labs.

The newest layer expands the impact of concern about the environment. Now, government and NGOs began to exert pressure in terms of extended producer responsibility (EPR) laws. Safety and tracking also gain importance as issues. An NGO named "As You Sow" puts a focus on corporate responsibility; in 2011 it put out guidelines to assess potential safety risks of nanomaterials in food packaging. When you deal with this category of influences, keep this in mind: Be transparent. Deal openly with NGO's and watchdog groups because they will bite if they think you are not being fully forthcoming. For a look at some guiding principals on doing that, read about the communication strategy outlined in Chapter 8.

Green concerns also drive expansion into forestry, farming and extraction of materials. Certifiers now validate where packaging is on an eco or sustainable chart. Certification raises conflict of its own. What is the best form of certifier: Industry-supported programs and independent third-party certifiers?

That's a quick tour through little more than a decade. In that time, we've seen packaging management go from working primarily with the organizations next to it on a value chain to a wide scope of concerns on the value web.

What's the effect of influencers?

How can that overview help packaging move into the future? First, by seeing how the new influencers will change business and packaging. The foremost impact is the move toward holistic views of packaging, manufacturing and business models. The value web serves to embed several considerations that weren't even "on the radar" a decade ago. Green thinking is now an integral part of the process among thought leaders. What we're calling holistic packaging becomes embedded in the development process. The value web also brings laws and regulations into the development process to a greater degree than the linear model did.

Here's just one example of how it translates to business practices. In December 2011, Kraft issued its "green" footprint report. Much of its content, and much of the work behind it, reflects corporate responsibility and recognition of questions raised by NGOs. The report assesses agriculture's impact on Kraft's carbon footprint, an example of just how far the value web has taken strategic corporate planning. While packaging is a small part of Kraft's carbon footprint, it is a high-visibility part; in its report, Kraft is quick to point out that packaging is down 100,000 metric tons in the 2005–2010 timeframe. The value web also embraces a global supply network. P&G, for example, sells its Olay brand of cosmetics in China with the same branding it sells in the United States. And, P&G has to deal with supply issues within a worldwide value web, including imposing Western-style quality control within the China market.

How does today's value web compare to the older value chain, and what impact does that have on material suppliers, converters and packagers? Here are some key issues that are currently playing out:

- The value web's span now reaches from farming and mining to package disposal, reuse and upcycling. On the supply side, recycled material sources will continue to gain in importance compared to traditional virgin material suppliers.
- Its cost/value structures keep changing, largely because of the volatility brought by its global reach.
- As packaging plays a larger role in company strategies, those companies face a more complex path through the web.
- Streamlining a company's relationship with the value web becomes a priority management goal. By minimizing the effect of low-priority actions, packaging professionals can focus on high-priority actions and make their responses more robust and agile in the situations that matter most.

The new priorities

Priorities will continue to change, and here are the directions major influencers in the value web will have in the near term. Figure 4.4 shows a how the priorities will change, relative to each other.

More consumer impact. With e-everything, they will exercise purchasing decisions from a position of greater knowledge. That means businesses have to address them with communication tools, including packaging. To do that, packaging will need to take more responsibility in translating market research and insights into the parts of the value web that deliver solutions.

Private brands. Chapter 3 details why and how retailers leverage them, and they are a growing tool for retailers in the future. Channels such as dollar stores are natural fits for private brands. Their management investment in packaging will grow—either by adding packaging staff or through third-party packaging developers.

Contract packaging and manufacturing. These companies grow and respond to private brands, retail uncertainty and innovation. They also address cost-cutting efforts. Their importance will grow for both national brand owners and retail managers as private brands find a new level in retailing. Contract manufactur-

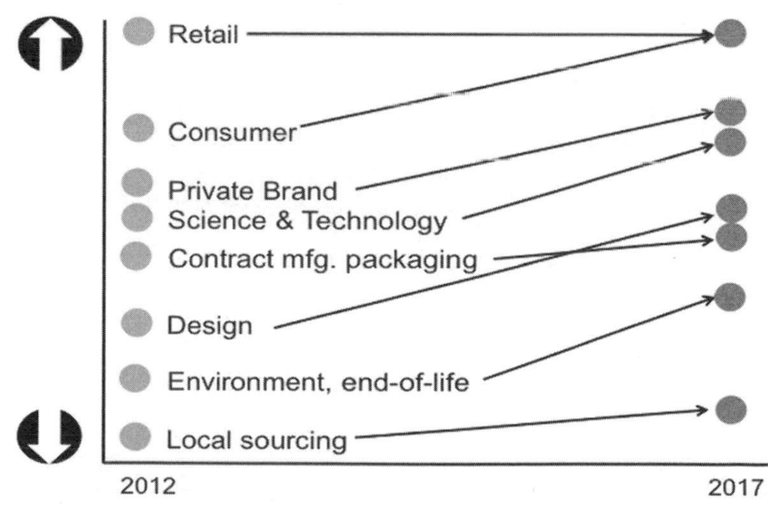

Figure 4.4. Component change.

ers and packagers have to deliver consistent quality so brand owners can build equity; the ability of brand owners to get production and packaging consistency is critical to success. There is already a solid base of contract packagers and manufacturers in place.

Packaging design. Each of the functions above relies heavily on design as an implementation tool. For those of us who have been around long enough to equate "design" with graphics, we need to keep telling ourselves that the function is much wider that that. It is structure, it is shape, and it is more. The idea of holistic packaging design that we outline in Chapter 5 will become the way we do things.

Environmental and sustainability. These factors continue to grow with particular emphasis on end-of-life issues. One trend we're seeing in 2012 is the work to bring non-bottle rigid containers into the recycling fold. Containers such as thermoforms are now seen as too valuable to be consigned to landfills.

Going global and going local, at the same time. Global strategies will remain dominant; they can reduce cycle time and speed commercialization. Tooling and samples can be created more economically in Asia, and it takes concerted management attention to work right. Yet, an emphasis is increasing on local sourcing; it can reduce the cost of fuel as well as carbon footprint. The concern expressed by economists that huge trade imbalances are a cause for concern will also drive some decisions toward local sourcing.

Retail. It is in chaos now. Just one example: Internet packaging has emerged as a discrete channel and is coming with its own requirements and those are changing. Other factors detailed in Chapter 3 illustrate the chaos. We believe that as chaos becomes the norm, the amount of effort to keep up with it will decrease.

IT TAKES A TEAM TO DEAL WITH THE WEB

The packaging value web is complex, and it puts a burden on its participants. If cross-functional teams for packaging made sense a decade ago, then they are mandatory today. Packaging has to either get or supply information to multiple functions throughout the packaging value web. It takes teams to do that effectively, and they include both people within an organization and from outside.

We spoke with Phil McKiernan, a vice president of Packaging Technology Integrated Solutions. One of Phil's areas of expertise is using teams to increase efficiency; he uses the graphic in Figure 4.5 to illustrate the packaging team's scope. It reinforces the idea that packaging is a strategic function and an enabler across all the functions within a packaged goods business. "We've got a decade of consulting in packaging management, and we've benchmarked more than 100 organizations with packaging functions," McKiernan explains. "These touch points may be inside or outside the organization, and each is critical to the idea of holistic packaging development."

One of the ways to interpret this graphic is to realize that not everyone has the same priorities. Businesses reward different functions differently, and each begins to develop its own perspective. One of the most frequent cited case stories of how that happens involves procurement (or as it is still labeled on some corporate org charts, purchasing). Incentives often focus on price

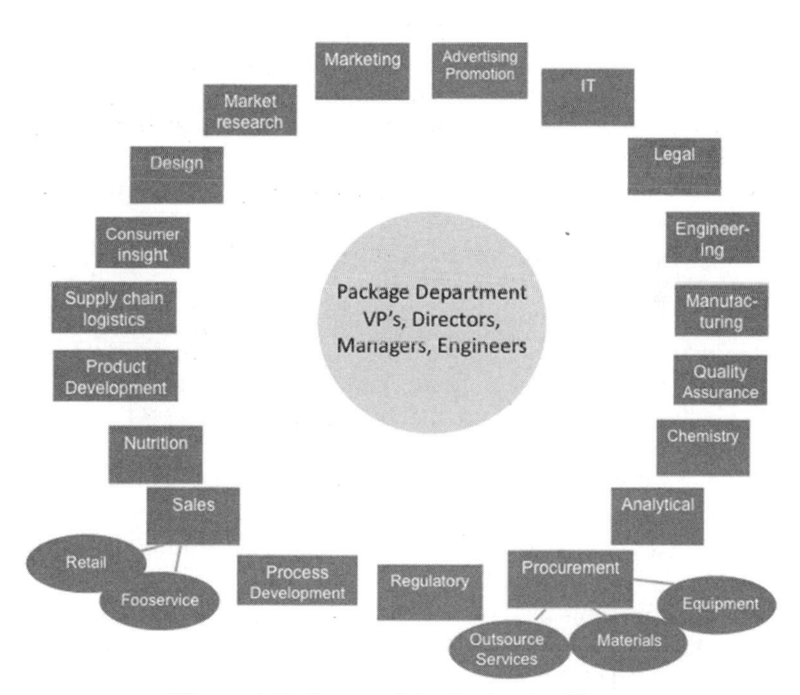

Figure 4.5. Scope of the Packaging Team.

performance for comparable products and services. That can diminish the incentive for innovation because performance can't be measured for that kind of change; that may be in conflict with packaging when it devises a discontinuous change that alters the package and end product.

Another difference in perception can involve the packaging/marketing link, and it is the job of packaging to translate from marketing's jargon and perspective to move the big picture project forward. When packaging is just a technical support function, the tendency among its professionals is to speak in terms of properties of materials. That's concepts such as oxygen transmission rate; technical people equate that with longer shelf life for many products—foods in particular. Marketers, on the other hand, think in terms of consumer benefits such as "freshness"; they also see opportunity in longer shelf life to extend distribution channels. Packaging's key role, says McKiernan, is to link the common interests, even where the jargon differs. Here are a few other key areas where the role of translator enables marketplace success.

Market research. The consumer insights and brand positioning issues we detailed in Chapter 2 are from the perspective of the market researchers. Packaging's role is to match those insights with solutions that are often technical in nature. Packaging's role also means putting those need/solution fits into language and communications that senior management understands as they make resource commitments.

Engineering and manufacturing. Historically, there have been good connections with packaging; the technical side of packaging development often parallels the thinking of engineering and manufacturing. The machine-to-package interface is critical, too. As electronic controls make machinery flexible, it becomes more a matter of give-and-take between machine requirements and the package's configuration. Cost control, operating efficiency and repeatability are other topics high on engineering and manufacturing's priority lists. If you go back to the value formula in Chapter 1, the relationship with engineering and manufacturing puts a heavy emphasis on lowering the cost. Yet, that is sometimes in conflict with marketing, where the emphasis can be on raising the benefits and experience.

Despite good rapport between packaging and engineering and

manufacturing, this link is often totally missed in the innovation process. Sometimes, the link is established in a project's final stages; management sees engineering and manufacturing in light of their historic focus on improving process efficiencies, increasing capacity, and replacing out-dated and depreciated assets. They have not been included in the front end of new product and package innovation efforts. Yet, engineering and operations groups can help early in the packaging development cycle by identifying the "best" way to develop and commercialize packaging innovation. Understand that the communication needs to be a dialogue—this group is looking for "the package" that they will design a machine and line to produce. But, in response, the developers are looking for new capabilities and ways to shape a concept/design to optimize for better efficiencies and lower cost. Look for equipment (engineering)/ operations and equipment suppliers to play a more integral and collaborative role in the packaging innovation process in the future.

The link with engineering, manufacturing and quality is so important that we also got input from Tim Brown. He's a PTIS manager with extensive work on integrating packaging development into engineering and operations. His basic message: collaboration is key. Make those functions an integral and up-front part of the development team to assure that the packaging materials will work within the ranges the machinery is designed to handle. "You have to understand and balance the capabilities of the equipment with the tolerances of the materials, and establish realistic finished good quality specifications," Brown explains. "Otherwise, you can expect to have missed expectations and more difficulties with line start-up and commissioning."

Brown suggests this strategy: Define and validate the operating windows of the various pieces of machinery that make up the entire packaging line. Then, establish performance-based material specifications that are within the packaging equipment's operating windows. The key is to have a wider tolerance in the packaging machine's operating window and a narrow tolerance with the material. Be sure the material's tolerances are within the equipment's operating ranges. That delivers enough of a "gap" between the two to help assure the material and machinery run consistently together and produce a package that meets expectations. Here's an example: a printed film may have a print tolerance of ±0.4 mm and Form-Fill-Seal machine may have a cut-to-registration tolerance of ±5.0 mm. The material print tolerance is significantly narrower and within the window of the cut-to-registration tolerance of the machine. The finished quality specification would identify an acceptable package cut-to-registration of ±5.0 mm and the line can meet expectations.

Finally, Brown talks about the critical importance of on-going training for packaging line operators, maintenance staff and quality control personnel on the operation of the equipment, the specifications of the materials and the interface of the materials with the equipment. It is training, Brown says, that needs to be done on a recurring basis, more than a session or two when the packaging equipment is first commissioned. "The on-going collaboration with engineering, operations and quality is to be sure the material and the equipment work together, and that the team understands how to identify and solve issues as they arise during normal operation," Brown emphasizes. Far too often, packaging lines run inefficiently as several changes are made to the equipment throughout production in an attempt to correct a problem without truly understanding the root cause and correctly implementing a corrective action.

Design. The role of design expands as the concept of holistic packaging development grows. We like to use the term Holistic Packaging Design system, and we will go in depth in Chapter 6. Within this scenario, it's easy to see how components such as shape and structural design gain importance, and the packaging function has to integrate the value of those components to operational and engineering activities. The role is to effectively communicate the complexity to designers who may not have expertise in engineering and manufacturing. The communications role also stretches to transport packaging and distribution channels. Packaging's role is to open options. It is another reason packaging needs to be in product development process early, so those options remain on the table and aren't removed simply because there's not enough time to execute them.

Marketing. Marketers may see risk in time-to-market issues. Different companies have different risk DNAs, and packaging needs to know what the tolerance is to make reasonable recommendations that affect time-to-market.

Procurement. Material sourcing, intellectual property control, and understanding technical, financial and consumer needs are concepts procurement has to embrace to do its job well. Logistics models and strategic supplier strategies are other key activities. The ability of packaging to both "talk" and to "listen" within interchanges with procurement increases probabilities of success.

Insights on commodity costs. Whether the expertise lies in packaging, procurement or some other function, forward looking estimates on commodity costs suggest long-term packaging strategies. The World Bank sees non-energy commodity prices edging downward in 2012 and through much of the next decade. An exception with an impact on packaging is aluminum: the World Bank sees its cost edging up in 2012 and beyond. The ability to get deeper than general forecasts like the World Bank's is a key to packaging success. We've seen run-ups in commodity prices in the U.S. based on how natural disasters impact petroleum feed stocks and production capacities. We've seen them based on global economic changes and political unrest. Being "pipelined" into information for the commodities that drive your packaging costs can be a valuable use of your networking time and could lead to material decisions based on long-term availability and pricing scenarios.

Gains from the systems approach. Most of the low-hanging cost reduction fruit has been picked and the savings in the future are going to depend on systems approaches. The way to be agile in adopting those systems is to have the value web that keeps you attuned to the network. Often, systems savings mean going beyond traditional packaging thinking. Developing packaging to maximize transportation resources and fuel is a prime example of doing that; doing it well means insights on current logistics thinking and practices.

STRATEGIC SOURCING

The connection between packaging and sourcing is so critical that it deserves special attention. In the best situations, sourcing and packaging work so closely that they work in synergy. At the other extreme, they compete, each working in a silo with a tactical mindset. And, there is a range of relationships that fall somewhere in between. The link between these functions can mean success or failure, and if you take a close look at packaging and sourcing, you will see parallels that can only be better when they work in synergy. That is particularly true with issues relating to innovation and packaging development. For an insight from the sourcing side, we talked with Marc Campbell, Senior Director of Strategic Sourcing at HAVI Global Solutions. The company handles package sourcing in a number of areas and looks in terms of a holistic strategy to deliver value. What it does underscores the need to encompass the entire value chain from raw materials to disposal of the packaging. Here are some of the key points from Campbell's perspective.

From a sourcing perspective, one of HAVI's efforts is to dampen price volatility and help deliver more consistent per-unit costs. That takes an expertise in economic conditions and commodity prices. Pre-buys and sophisticated hedging are some of the tools it uses to take some of the volatility out of packaging. It involves proactive work and sophisticated hedging, Campbell says. And it should involve techniques to measure performance.

We talked with Campbell to look at sourcing in terms of the Integrated Value Web, and what intrigued us was his organization's approach to strategic sourcing initiatives. Within the HAVI jargon, that range runs from pricing agreements that are tactical and straightforward. Figure 4.6 shows how they fit along a continuum. Perhaps a handful of arrangements fall within the Supplier alliance category, yet these offer the greatest potential benefits for both sides. When looking for those kinds of alliances, Campbell says there are four criteria in building a relationship.

The supplier needs to be integrated back to raw materials. This acknowledges the goal of dampening volatile prices, and HAVI asks for the kind of transparency by these suppliers so it can effectively monitor the way commodity costs contribute to overall costs. Even more, HAVI looks for alliance partners who can sell commodities to other suppliers in the value web. "We may have reasons, such as diversity, for bringing other converters into our system."

The alliance partner also has to be willing to put engineering and product development personnel within HAVI facilities. That supports the goal of having co-development agreements with the suppliers. "With the right suppliers, we're willing to share the

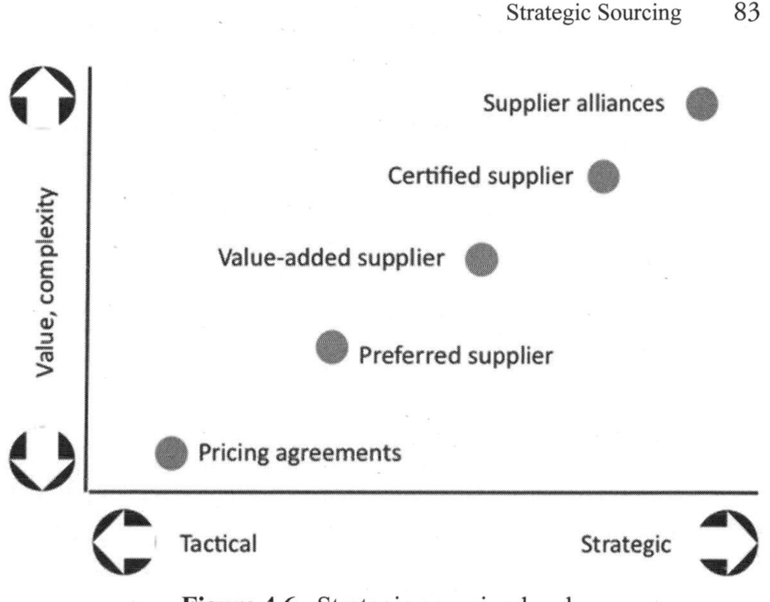

Figure 4.6. Strategic sourcing levels.

costs and the rights to intellectual property that comes out of the process," Campbell continues. As a final point, HAVI is willing to invest in building supplier plants to produce an item. That step also shows the commitment needed to have this kind of reach. The investment and planning can include where to site plants to serve the customers best, the kind of technology platforms on which to build capacity and even down to the equipment manufacturers that can best fit the requirements.

Strategic sourcing—from the HAVI perspective—embraces quality measures that go beyond the package's ability to meet its specifications. They build in audits for safety and sanitation, and assessment of packaging supplier HACCP programs. Running mock recovery drills is another. The effort even includes third-party audits of HAVI itself, particularly in terms of performance against category price against industry competitive sets.

Even further, the programs echo the concept that the value web includes a lot of influencers who are a part of the process. Social responsibility is part of the sourcing effort. HAVI looks at accountability of suppliers, and the suppliers' suppliers—its all part of protecting the client's brands. Another question to ask is: How do suppliers treat their employees? What are the suppli-

Strategic sourcing for strategic packaging

In Canada, a foodservice company was adding premium roast coffee to its menu, and it needed cups to convey the brand positioning. An emphasis on eco-friendly packaging within Canada meant the cups had to be made from renewable materials, and that precluded existing paper/plastic technologies, even though they kept coffee hot and carried high-quality graphics.

Technology scouting became part of the process. The sourcing organization—HAVI Global Solutions—found a European, paper-based solution. In another step parallel to packaging development, HAVI's staff came up with testing protocols for the new cups. It was part of the collaborative process with the supply partner.

The answer was a double wallpaper cup in multiple sizes. The finish on the exterior ply delivered high-quality graphics. It also offered a promotional option of an integrated loyalty card. It supported a program that saw coffee sales double in three years.

ers' ethics? Supplier assessment tools include audits of consumer complaints on packaging. When HAVI has complaints, it uses track-and-trace mechanisms to identify who the supplier was, what plant made the item and even when it was made.

CONTRACT PACKAGING—SPECIALIZED STRATEGIC SOURCING

Contract packaging and manufacturing gives brand owners one answer to volatility, uncertainty and ambiguity. It is particularly important as the number of packaging formats gets wider and brand owners look for more distribution channels through which to market their brands. A big plus for contract packaging and manufacturing is better speed to market—rather than starting to engineer new packaging lines, brand owners can shorten the lead time by using capacity already in place with a contract packager.

The universe of contract packagers is global—more than 400 in the U.S. alone and a quick search in Europe turns up more than 600. In many ways, China is one huge contract manufacturing and packaging zone, relying on lower labor costs to gain business from global customers. The process offers brand owners and retailers speed to market, increased production capacity, and the ability to handle package complexity. Contract packaging also can reduce capital costs, maximize equipment versatility and provide "insurance" against volatile product life spans. Contract packagers fall roughly into three basic types of operations:

- *Primary packagers* put foods, beverages, household chemicals, pharmaceuticals and a range of products in their primary containers. The range of packaging formats they can handle is a central benefit to brand owners; it allows them to venture into new formats without making a capital investment. Often, contract packagers also manufacture products to the brand owner's specifications.
- *Secondary packagers* are sometimes called assemblers. They combine primary packages into multipacks, often for promotions. Think of a brand owner introducing a new household cleaner by "piggybacking" a small container of the new product onto a standard-sized container of existing products. Or, think of a promotional "two-for" offer with two primary packages bundled together.
- *Display packagers* have mushroomed as brand owners create special promotions to get sales lift at retail. This kind of packager builds pre-loaded displays, and even free-standing display pallets. The emphasis is on shelf impact that drives impulse sales. The contract packager can add expertise the brand owner may not have on staff. Labor costs are lower than the brand owner's costs.

From our experience, brand owners or retailers who want to contract package their private brands need to go through two decision processes. First, a matrix deciding how the business risks fit their corporate culture and objectives. The second is whether the costs for contract packaging are acceptable or whether it makes more sense to set up the packaging operation in-house. According to the report "Contract Packaging: Strategic Opportunities & Profit Potential," here are some of the business risks that impact the contract packaging decision:

- Does the new package differ significantly from existing products/packaging? Will the product/package fit on current packaging lines? Would you trade off product/package benefits to fit into existing packaging capabilities?

- ROI/payback hurdles: How big is the investment and what is the likelihood of reaching return/payback goals?
- Maturity of product category: Will the maturity of the product category affect potential success? Does it affect the risk of failure, either with the risks of a new category, or the risks that competitors may respond in an established category?
- Estimated product lifespan: This factor may be particularly significant for line extensions or for promotional offerings.
- Need for confidentiality: Contract packagers sign confidentiality agreements, but breaches in confidentiality do occur. New developments have been compromised when a prospective client (and a competitor on the new development) looked behind a curtain while doing facilities evaluation. If the need for confidentiality is very high, a contract packager may not be the right answer.
- Product/package technology: How big a jump is it for all the participants in the packaging value web? What is the likelihood that the technology can be implemented without time setbacks and process modifications?

For a detailed perspective on cost analysis, see the Appendix and Resources.

FINDING TECHNOLOGY'S OPPORTUNITIES

Scan the packaging value web for opportunity, and you will find few areas with more promise than emerging science and technology. It is an area that can deliver disruptive solutions—new packaging formats that change product categories. The value they deliver raises market impact, consumer benefits and margins. Often, the technologies can do it at a price competitive with methods that don't have the same consumer impact. Consider laundry liquids. Concentrates gave a significant yet incremental increase in value. Now, Tide Pods and other brands relying on similar technology are disruptive solutions; they are based on technology that takes an entirely new tack in the category. It is just one recent example of using technology to deliver value.

Where are the most relevant answers emerging? We talked with Ross Lee, Ph.D. He is a Senior Associate with Packaging Technology Integrated Solutions. Lee heads up the PTIS Science & Technology Project; it is a high-level effort to identify packaging's top needs and the technologies to address them. He isn't the slightest bit vague about how he sees science and technology influencing packaging's future. His observations have a solid base; he himself has some 20 patents and publications, and he's been through the inner working of the intellectual property process. In the broadest view, he sees big science and technology addressing these trends:

- *Climate.* A broad strategy has already emerged to replace fossil fuels and feed stocks with renewable energy sources and materials. That we have the term "carbon footprint" in our packaging vocabulary confirms this trend's importance.
- *Sustainability and green packaging.* It goes beyond materials technology and embraces the triple bottom line of people, environment and profits. Its impact has grown so fast because it works for environmentalists and it works for chief financial officers. It reduces waste and problem materials; at the same time it cuts process inefficiencies (think costs) in material, converting and packaging. Lee believes that sustainable packaging technology's benefit lies in both immediate and long-range solutions.
- *Connected world.* We probed the e-everything consumer in Chapters 2 and 3. It is a group of consumers that wants knowledge anywhere at anytime. The Internet, social media and the package itself can provide those routes, giving consumers the knowledge they want. That same "connected technology" also gives packagers and marketers the tools to provide the connections consumers want.

With input from people like Ross Lee and a team of experts, our "take" on technology spans from "quick fixes" that work today to long-term directions that may give us a packaging industry with new foundations in the future.

Materials

From a strategic viewpoint, here are some of the areas where those technologies need to focus to address the bigger picture.

A high-priority area is extended shelf life for food products; one driver of that priority is developing countries as they move market societies. Refrigeration is a "given" in the United States, yet the energy costs involved in the cold chain are a significant part of a product's carbon footprint. In other parts of the world, the need for ambient temperature distribution isn't a convenience—it is critical to developing societies. As economies such as Sub-Saharan Africa grow, the need for longer shelf life in non-refrigerated foods can support that growth.

Climate concerns and sustainability have pushed the growth of bio-based polymers. The action in the next few years is in finding the right "fit" for those polymers. Right now, these materials are niche products with potential to become mainstream as quantities increase and prices come down. We're going to see a wider range of high-performance materials produced from renewable feed stocks. The key is building performance into the packaging materials with routes such as coatings or mixing of materials. Yet, that approach may face some problems, especially in the face of thinking that suggests one-material structures to simplify recycling. Here's where the value web excels as a framework to assess options—raw material suppliers and converters can work with brand owners, retailers and recycling organizations to investigate performance for specific applications.

Also expect niche market application of compostable and biodegradable packaging. This is leading edge technology. Work on biodegradable coatings for paper aims toward making those packages fully compostable.

Nanotechnology

In many applications, nanotechnology is a key to imparting properties that give packaging's base materials the functionality they need. The biggest roles nanotechnology can address are adding strength, barrier and special effects. This technology will help drive "smart" packages. Work at Rutgers and the University of Connecticut helped develop color-changing nano-sensors that could signal spoilage in foods. Research also indicates that nanotechnology could also enable electronic information dissemination with signal transmission from packages. Nanotechnology is not without risk. Researchers are assessing what some call "a huge gap" in knowledge about the effects of nanoparticles. Ross Lee points to a risk assessment framework that first involves detailing what nanomaterial is being used, the application, and then asking the question, "Is it inherently safe"? The next step is to develop a lifecycle profile and determine a risk profile.

Information technology

We've looked for ways to drive home information technology's impact, and we came across this from CNN: Peter Diamandis is a channel of innovation, perhaps best known for offering the prize that launched non-government space flight. He says that today a tribesman in Kenya who has a smart phone has more information at his fingertips that the U.S. President did as few as 15 years ago! It's an amazing leap in communications; we're not sure of the connection, but Africa is certainly seen as a probable crucible of world economic growth.

How do advances like this involve packaging? The consumer's ability—through smart phones—to read 1D and 2D codes makes the package a catalyst and doorway to a tremendous amount of information. The utility of long-existing codes such as UPC, EAS and ISBN barcodes now extends beyond the retailer to the consumer. Already, half the mobile phones in the U.S. are "smart" phones with the ability to read those codes. The evolution of smart phones and phone-readable codes blunts one of the drivers for RFID technology. The need to provide robust information to consumers can now be accomplished through a printed code and the Internet rather than a RFID chip with coded data. Despite all the enthusiasm of the mid-'00s, the cost of RFID chips has not come down below 5-cents per chip, a number that may represent the upper cost packagers could pay to utilize RFID on individual packages.

BE A TECHNOLOGY SCOUT, KNOW WHERE TO LOOK

The biggest challenge is to find the right technology for you without being overwhelmed by the options out there. Here are some ways we think packaging managers can sort and assess technology in the most efficient manner.

The first step is prioritizing relevant technologies, and one way to do that is to ask of any technology: "What is the potential business impact and is it big enough to justify the investment in time and resources?" Here's where the holistic approach matters; if you know your own company's business goals and strategies, you are in a better position to answer that question. You need the same insights to answer the second question: "Is it feasible to implement the new science or technology within my company?" The answer depends on your innovation timeline. Are you working on a short-term project and do you need a technology that is ready now? Insights into the products suppliers offer can be the most relevant route here. On the other hand, if you are looking long-term, an approach that is at a university or a technology incubator may be worth tracking. Another question to ask: Does your organization have the infrastructure and capability to use a new technology? For example, are you practiced at the testing and research capabilities needed for a specific technology? We think of packaging gaffes like the bio-based water bottle resin with such a high water vapor transmission rate that bottles visibly "paneled" on store shelves. It was a performance issue that an organization committed to the right testing should have spotted.

We like the term "technology scouting." The scope of technology scouting depends on your organization's goals. Some may want a broad search across many technologies; other may have a single niche they need to know in-depth. As a guide, here are four places to look to gain the insights on how technology can address your needs.

Technologies developed in the past. These often come from industries beyond packaging or in packaging niches where they meet specific needs, yet have minimal visibility. A classic example that changed an entire product category's packaging is the Starkist tuna retort pouch that hit the market in the early 2000s. It answered consumer needs for convenience and better tasting tuna. The retort pouch had the capacity to do that, and it offered graphic capabilities to attract attention on store shelves. At that time, the technology was at least 25 years old in the U.S., and had even earlier roots in Japan and Europe. New market conditions made it the right answer in a different product category. You can even find old technologies within your own "four walls." Check the ideas at your company that didn't work in the past; another source is technologies that worked once but disappeared as market conditions changed. New market conditions may establish a base to bring the technology back.

Technology solutions under development. What are others developing now? Don't limit your search to those technologies under development in packaging. Look in areas along the value chain where developments may impact packaging. In 2011, major areas were electronics and data management. One example involves technologies in vending where concepts such as "touch screen" interfaces with consumers may open opportunities for new packaging formats. As you look for technology in other areas, you may find technology developers that are looking for partners to offset development costs.

Future technologies. By definition, you can't know what to look for in future technologies. Here, the focus is not on what, but on who. Build a list of organizations and people who are on the leading edge in specific fields. When your needs stretch into that area, you have contacts to extend your insights.

Bundled technology components solutions. Sometimes, one technology alone may not give you a solution. It may be necessary to identify additional technology components to deliver a solution. Back to our example in vending: Touch-screen technology may open new product opportunities, but you may need in-line quality control technologies to deliver them.

Packaging is so integrated into the value web that some talk of packaging as a social function. They mean that its effectiveness is based on being able to reach out and gain insights from the right people in the right parts of the value web. Our experience says there is truth in that, especially from the perspective of the number of outside influencers. Yet, packagers also have to keep a foot in the technology arena. Both are part of the formula for success in the future.

Packaging value web action agenda

The integrated value web is global

That includes all facets of packaging—material suppliers,

equipment suppliers, university think tanks, influencers, governments and others.

Packaging Action: If you are not leveraging the value web's global players, you are not getting all the value the web offers.

The number of influencers continues to grow

Many of the new influencers are organizations that go beyond the supply chain concept. They impact packaging by putting conditions around its use. Consider European extended producer responsibility organizations as an example.

Packaging Action: You need to collaborate with them. It may be in promotional activities such as working with the World Wildlife Federation. It may be educational activities with any number of eco-focused organizations.

Teams are the best practice to get the most from the value web

Packaging Action: The integrated value web is simply more complex than what the value chain was just a decade ago. You need to bring a number of disciplines into the process.

Systems thinking has to dominate

Rarely does an effective answer rely on just one part of the value web. You gain the edge in bundling solutions.

Packaging Action: Keep your contacts open to gain insights on all factors that impact your packaging.

Set a technology focus for your specific needs

Being effective requires a focus that depends on your company's specific needs.

Packaging Action: Be a technology scout to have knowledge of what's out there to solve your challenges.

Packaging and procurement need to work in synergy

If there is any place where an organization can't afford silos, it is between packaging and procurement.

Packaging Action: Work to build value web relationships in synergy with procurement's efforts.

5 || *Design: A Soul Issue that has to be Holistic*

> ***A scenario . . . In the USA.*** Package design is the "soul" of Method Home Products. Coupled with eco-friendly products, the brand carved a niche in the sea of sameness that was home cleaning products. The brand's breakthrough was packaging design that brought aesthetics to the kitchen sink. Sporting the colors of cucumber, lavender and mandarin orange, the packaging looked more like a section in a candy store than a cleaning products shelf in a traditional superstore. Analysts put Method's annual sales in 2011 at about the $100 million mark. And holistic design with soul is a key driver.

"Look at the most successful packaging designs today—the ones that sell brands. There's a different sense about them than just a few years ago. Going in, the designer draws on a lot wider range of inputs than before. And when a concept become a package, it reaches out and becomes part of the consumer experience. It is holistic." That's the view of Paul Castledine, Chairman & Chief Creative Officer of Boxer Creative. Castledine is one of our colleagues at the UK-based design agency. Boxer's clients are around the world, and we spent some time talking with him to get a global perspective of where today's packaging design is, and where it is going.

"Packaging design goes way beyond the visual, brand and retail impacts. We hear about the First Moment of Truth—where shelf impact is the key objective," Castledine continues. "That remains important, yet today design has to go way beyond it. Packaging has to become part of the consumer's life and experience. Look at Kleenex; it uses design to embrace the seasons, and it increases its margins when it does that. Here's another one—a box printed in Pantone 1837 is not just a box. It is a Tiffany box and all the emotion that goes with that. We're looking at examples where design is a soul issue, and we're going to see an explosion of that in the next decade. Yet, the idea of embracing wider inputs makes the process holistic, too. You need that."

Fred Collopy is another noted designer we've talked with and he says, "Managers don't just have to hire great designers. They have to be great designers." And business guru Tom Peters is part of the chorus, too. He's said that one of the top mistakes in business is to treat design as a surface issue when it is a soul issue; it is not something to "tidy up" the mess at the end of the process.

We're not sure that within this book we can turn solid managers into great designers, but we are going to give you some tools you can use to increase your design savvy. These tools can take you a long way on the journey to be effective in integrating packaging design into package development.

HOW TO LEVERAGE PACKAGING DESIGN

Here's the core challenge that packaging managers face when they deal with the design piece of the holistic packaging process: How can you best capture design's impact within the process? We think a key is to realize that design's scope goes well beyond aesthetics. When packaging is a strategic tool, design has to encompass sustainability, operations, business objectives, innovation and universal design. At Packaging Technology Integrated Solutions, the term is Holistic Packaging Design TM system. It means the design phase includes all the components that affect the package's final configuration. Packaging management's task is to reach out and make design an integral part of package development. In most cases, that means elevating its presence in the development process. At the same time, designers have to reach out and use more tools to meet complex demands.

We see these major benefits by adopting this approach.

Reduced total cost of ownership. We touch on total cost of ownership in Chapter 9, and the design phase is one place to have a significant impact on it. Not only are you determining the materials costs, you are determining operational and logistics costs.

Setting up the base for consumer experience. Some products gain a consumer following because of the experience in using the product. Method Home Products is certainly one. Apple is another. Any number of gourmet frozen meals are also experiential. By taking a total approach to design, your chance of conveying that experience to consumers expands.

Proprietary, ownable elements. The cap on Dean's Chug milk packages has a design patent. It is a product of a holistic design program, and it has conveyed a message about the brand that has been effective for 15 years.

We can best define the holistic process by examining the components it includes. What you see in Figure 5.1 are all the components of holistic design; they go beyond the design brief packaging developers have been sending to agencies for years. Even for those elements that were within the traditional design brief, the process now goes into greater depth.

Consumer. We detailed a lot of what impacts the consumer in Chapters 2 and 3. Yet, it goes further. Today's world is on sensory overload and the way to gain attention is to add more sensory reinforcement. It can mean tactile features such as soft-touch inks and resins; it can be holograms, and even "talking" packages. The answer will depend on the package's target consumer, distribution and end-use environment. At the outset, ask the question, "What sensory element do we need to have?". Design

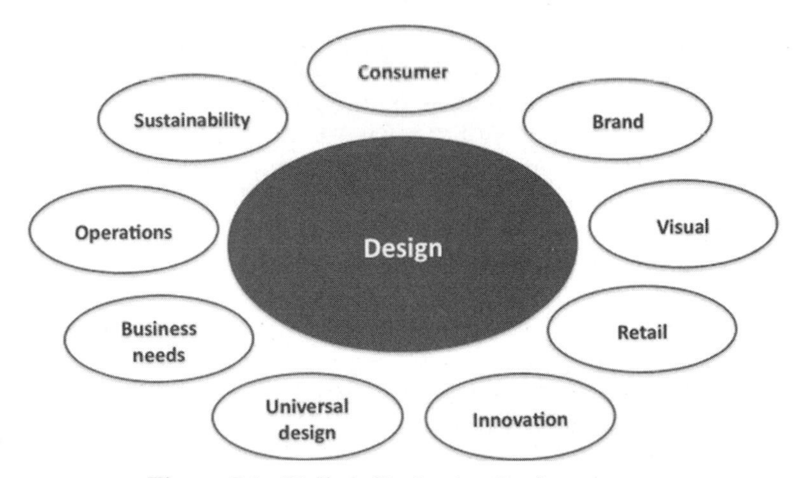

Figure 5.1. Holistic Packaging Design system.

goals for the consumer also embrace experiential factors. Using Method cleaning products as an example again, they changed the consumer's experience with the product by taking it from under the sink and putting the package onto the countertop. Method packages do that by delivering an "AHA!"

Brand. Any number of valuation systems will detail the worth of a brand. A number of the high-ranking brands have packaging as a key component in their strategy. Coca-Cola, Tesco, Gillette, and Colgate are among the top international brands. Brand value will vary by product category and global region, but strong brands will gain a premium in the marketplace. In design, projecting the brand value is often a top item in a design brief. This is true, too, in developing markets where consumers see a value in an established brand versus emerging local brands that may have quality problems.

Visual. Color and shape are the foremost tools here, with shape often as important as color in gaining recognition. The rise of shape within the visual hierarchy is a factor that pushes us toward holistic design. Now, designers need to be more versed in materials and production techniques, and packaging managers have to be sure the designers know those parameters. Packaging's failures include a number of stories where design didn't understand production limits—for example, bottles that broke because a shoulder was too "square."

Retail or distribution environment. The ability to have impact at the First Moment of Truth is a make-or-break factor. A packaging design may look great on a designer's monitor, but how does it work in a shelf set in a real store? Or, for business-to-business product, how does it work at the First Moment of Truth there? If it is a medical device, for example, does it provide proper identification at the point of use? Can the person who first unpacks it open it easily?

Innovation. Innovation is bringing the right technology together with the right marketplace needs, and doing it at the right cost. Holistic design helps link these components. Management's role here is to bring packaging value web players into the design process. That can fully integrate innovation's contribution in the design. The designer has to grasp an innovation's impact to be able to communicate its benefits. Our experience goes back to the first slider zippers where some packaging de-

Herbal Essences and holistic design

One brand gives us a unique look at holistic packaging design's impact versus a traditional packaging change. That brand is Herbal Essences, a hair care line that was repackaged in the late '90s and then repositioned in 2006. Clairol owned the brand in the late '90s, and it faced sagging sales. The answer was a solution that modernized the brand by leaning heavily on packaging's ability to project "newness" and "added value." The package was a clear plastic bottle with a see-through front label. It was a leading edge approach then, and offered the promise of shelf impact with a new look. The bottle was new and more contemporary, yet it stayed within the "apothecary style" of its predecessor. It showed the product through the label in what was then a dramatic splash of color. It was a traditional redesign that produced a spike in sales. Yet, five years later, when the brand went to P&G, reports called Herbal Essences an aging brand; some say the term "death spiral" appeared on memos.

Enter a P&G holistic process that first looked at the consumer. Clairol's target market was basically any woman who washed her hair. P&G's development team opted for a new niche—Millennials who were in their late teens to early 20s then and didn't have a hair care brand targeted at them. "Everything starts with the consumer" is a P&G mantra, and extensive research found a different kind of sexuality for Millennials—subtler than the Gen-X overt sexuality. The research drove the holistic design process, and the P&G team took just about three months to put together a plan that changed everything about the brand.

Where package color had been the main differentiator in the hair care section, Herbal Essences added shape. New bottles "nested" together with the curve of one fitting into the curve of the second bottle to support a sensual look. One held shampoo, and the other held conditioner. But, it was more than an exercise in aesthetics; the P&G team had been to "production school"—learning about package production and line operations. The "nesting" bottles are, in fact, the same bottle, with one flipped over onto its top. That simplified package production and line operations to reduce risk in implementing the changes. The rebranding appealed to the target market with names for variations that had 'attitude." One variety, for example, is the "Totally Twisted" collection for curls and waves. While P&G won't reveal sales data, reports put the repositioned Herbal Essences early growth rate into the high single digits.

signs actually "hid" the innovation; more savvy designers—on the other hand—found ways to highlight the innovation's benefit to consumers.

Universal Design. Especially in the parts of the world with aging populations, we need design that works for those with physical impairments. We explain it later in this chapter.

Business needs. One of the pillars of a systems approach is that packaging aligns with the mission, vision and objectives of a business. Design has to support the alignment. A key element in that is total cost of ownership, and we'll cover that in detail in Chapter 9.

Operations. Can production run it effectively? The archives of packaging failures are full of packages that did everything else well, except they couldn't run on production lines efficiently. That pushed per-unit costs beyond the margins built into the original planning. In the worst-case scenario, safety issues arose when production requirements and the package couldn't work together. The results of not bringing production in early can result in costly redesigns and setbacks in product launch dates.

Sustainability. We will cover sustainability in Chapter 6, but make sure the right questions are part of the brief. In particular, focus on end-of-life issues. As much as 80% of a package's carbon footprint can be determined at the design stage.

Meeting the CPGS' Challenges

It's easy to see design's impact in smaller organizations such as Method Home Products, but we wanted to get a look at how design principles play out in larger, successful consumer packaged goods companies. We talked with Peter Borowski—he's a senior design director with Kraft Foods who brings more than 20 years of agency and corporate design experience to the table. Peter follows four guiding principles to ensure design success.

Know your business challenge. "Design succeeds when it clearly and intuitively communicates your product message," Borowski says. "Design has to convey the product promise and benefit, in a simple way." To do this, designers need to understand the business challenge behind the product and brand. The measurement of packaging design's success is how quickly and effectively you can communicate "who I am and why I am right for you." Getting there is an exercise in partnering and aligning with a cross-functional team—marketing, design, R&D (Product and Structure), consumer insights, sales and the agencies.

Know your consumer. In Chapter 2, we made a huge point of knowing the consumer. Borowski underscores that it is mandatory that packaging designers leverage consumer insights. He puts it this simply, "If you don't know your consumer, you can't design to their needs".

Know your competition. How does their packaging look on-shelf and what are their communication strategies? Borowski's approach to understanding this involves both on line and in-store research. His suggestion is to spend some time up close and personal with the brands in the categories where you compete. "When you're shopping, spend more time in the aisle where your category is shelved. Go on line to do the same. Look for trends, then draw conclusions from what you see," he offers. "You may also see a brand encroaching into your territory. This could be a style of photography, illustration, typography or structure. It could just be color." He believes an assessment of the competition is critical to developing great design: Is it proprietary and ownable? "If you're not proprietary and ownable, you're 'me too,'" he says.

Know yourself. You have to know what your brand stands for. "With great brands, consumers should be able to easily recognize and describe the brand in one sentence," Borowski explains. "Consider Apple. The brand is consistent across all communication touch points—visually and verbally. Their packaging, advertising and in-store experience is seamless." A consumer identifies a brand by its equities—elements that provoke a positive and emotional experience. Borowski points to Maxwell House Coffee where the color blue and the tag line "Good to the last drop" are equities consumers play back when they talk about the brand.

Finally, Borowski says it's imperative—especially in a larger organization—that a designer be a good communicator and listener. "It's being able to work with marketers and understand their business needs," he offers. "And educating them about the benefits and value of good design. Consumers are simply more design savvy today. Target brought design to the masses at affordable prices. Apple leveraged design to take something complex and make it simple and intuitive. The level of design is simply higher, and designers have to deliver that just to be competitive and win at shelf."

Roadblocks

While a Holistic Packaging Design system is a best practice, organizational roadblocks get in the way. Here are some we see frequently in the packaging development processes:

Too late into the process. The packaging function and design are brought into the development process too late; the result is that some options that could have precluded problems simply aren't doable because of time constraints or conflicts with other criteria that have already been established. Here's a perspective. If we look at the traditional packaging development process, we find that packaging design often enters into the process relatively late. Figure 5.2 shows the process. Not uncommonly, design comes in at the product and marketing development stage. Yet, much of the emotional connection insights emerged much earlier. We're suggesting that design becomes part of the process at information mining and platform stage; this is where the "spark" of consumer needs emerges and where many of the emotional insights emerge. If we wait, the widest range of options is already "off the table" because of cost or time limitations. That can be particularly true for structural and functional packaging innovations.

For companies that use a form of the Stage Gate process, it may mean adding a Stage 0 to integrate packaging into the process. In a typical Stage Gate process, the conclusion of Stage 1 produces some estimate of product/package configuration; that in itself precludes some options. Adding a Stage 0 means sur-

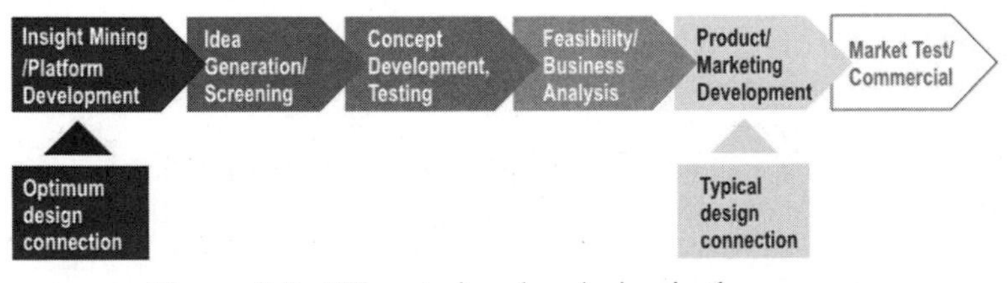

Figure 5.2. When to involve design in the process.

veying the product/package combination early so the scope of options expands.

No sustainability criteria. This happens less often as sustainability works its way into corporate DNA. Know that the packaging brief has to include end-of-life options such as recycling, disposal or reuse. Other factors in end-of-life criteria: Can the package be up-cycled (used in a product with a higher economic use)? Can it be deconstructed to recycle more easily? This a principle that emerged in the original sustainable concepts for products such as office furniture. In packaging, examples include the ability to separate plastic from paperboard components to get them into different waste streams.

Narrow focus. The systems approach says design needs to go beyond the primary package in a single distribution channel. Today, few products sell only through grocery channels. As the number of channels increase, so does package development complexity. If you sell through club stores, the role of secondary packaging multiplies. Software exist to tie together primary, secondary and tertiary packaging. Ask the question, "What are all the channels where this product could be sold?"

Teams. Cross-functional teams are a must. Covering all these bases can't be done by one corporate function; packaging can coordinate, but teams need to act on all the elements involved.

For a designer's slant on the process, we talked more with Paul Castledine. We talked, in particular, about how packaging managers can make the design process more effective. "Design has to balance the creative with the pragmatic," he explains, "It is building the emotional connection with the consumer—we use the terms 'desire' and 'love.' But we know that we have to connect that with all the pragmatic and rational components. I think packagers get the most out of design when they give designers all the nuances on the product—its consumer, distribution channels, production and the rest of the Holistic Packaging Design system. Then we can do the best to make the package be the brand and make an emotional connection with the consumer."

EMOTIONAL CONNECTIONS

Soul issues are emotional connections, and here's some research that sheds light on the ways consumers connect with products through packages. The research is from Packaging Technology Integrated Solutions, a division of HAVI Global Solutions. The proprietary work also involves partner GfK Group. It probed what motivates consumers as they buy packaged goods, and it identified emotional dimensions that are part of the buying experience. Packaging helps to deliver those connections, and it makes sense to get packaging into the insight process early when marketers first explore the emotional connections.

The research looked at four drivers of purchase; these drivers often emerge at the insight mining and platform development stage. They are:

- Safety
- Wellness
- Gratification/enjoyment
- Convenience

Within those drivers, the PTIS research went on to identify what are called the "dimensions". Each dimension is an attribute the consumer seeks under a driver.

Safety is a driver defined by this phrase: "It won't hurt me, my family or the planet." Consumers may respond to dimensions such as "pure," "integrity/trust" or "organic". Safety is not the primary driver of a purchasing decision, but it evokes an emotional reaction to the product/package combination. The safety connection can be a "tiebreaker" when all else is equal. We've seen the same condition from research on sustainability—the factor is not a primary decision-making criteria, but it can tip the scales if all else is equal. The PTIS-GfK research uncovers another connection related to this driver—a strong brand plays a major role in reassuring a shopper that a product is safe.

Wellness is a driver defined as: "It is really good for me." Consumers react to dimensions such as "holistic" or "fresh." This purchase driver shows how packaging can address a driver. Package transparency, the use of green or neutral color schemes and package resealability support the wellness connection. With this driver, too, brand plays a very strong role in consumer perception of whether a product delivers a wellness benefit.

Gratification/Enjoyment: "It makes me feel good. I'm worth it." Consumers have emotional and aspirational connections to this driver and they react to dimensions such as "premium" or "authentic/real." Common packaging tactics to signal gratification/enjoyment are bright, saturated colors and bold fonts.

Convenience: "It's so easy for me." This is the most packaging-driven driver and survey respondents were often able to articulate how certain packaging elements made the product/package more convenient. Convenience is highly visible and structural packaging elements are central in delivering it.

As we look at how packaging managers and designers can integrate emotional connections into the packaging process, we go back to Paul Castledine and his view of design. "Here's one thing I would suggest to all the packaging and design managers: You be the shopper and the end user of the product," Castledine says. "Advertising used to be the primary media. Today it is the store shelf and the computer screen. The rule of thumb is that 70% of purchase decisions are made in the store. Packaging managers and marketers can see the process better if they shop

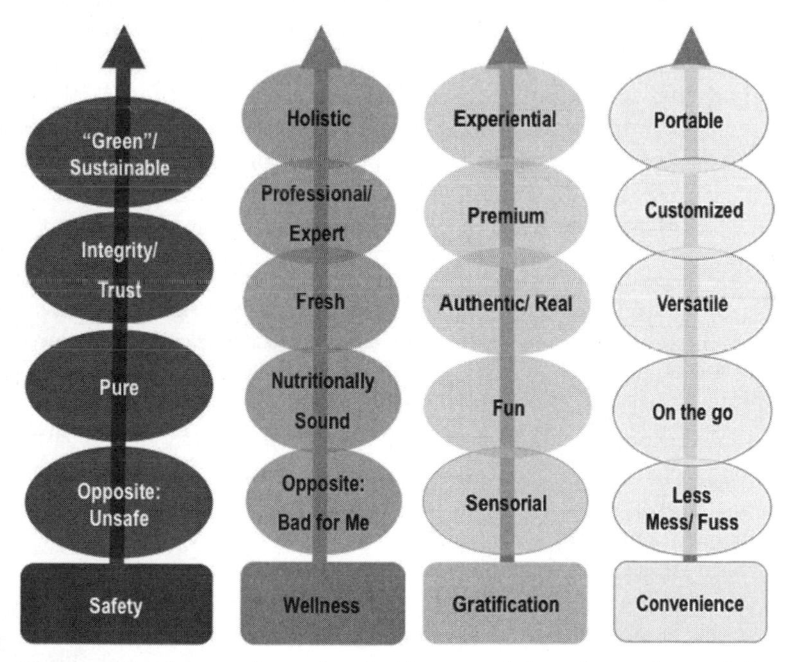

Figure 5.3. Array of emotional drivers and key dimensions within each driver.

and use the product. You begin to get a feel of the challenge. If you're the leading brand, think that you have to differentiate yourself from the competitors who want a bigger piece of the category. If you are a private brand or a follower, how can you grow in the category?"

GET THERE WITH THE PACKAGING BRIEF

The packaging design brief is the usual implementation tool. As design and packaging development have to do more, the brief's role has grown significantly. In best practices, it touches all the packaging solution drivers—graphics, shape, structure, sustainability, and operations. It touches all the functions in Figure 5.1 and becomes a tool to manage the entire project. Because of that, we simply call it the "packaging brief." It is the first place where all the factors that shape a solution come together. It becomes the strategic platform to guide package development. Yes, it has the traditional elements of a design brief—equities, consumer input, brand positioning. It also engages other priorities—sustainability, operations, end-of-life scenarios. It embraces all the elements in translating insights and visions into packaging that sells products. Done right, it is a solid business document that

Making package design a team decision

One case story involves a packager of maraschino cherries. The company wanted to boost sales, and its research revealed a significant niche for cherries used on ice cream sundaes; the target audience was kids and moms and the goal was to delight them with the product and package. The marketer had a structural designer come up with a "fun" jar that would capture the attention of both. Just before he signed off on having the molds made, he showed the jar to his operations staff whose job was to fill, cap and label the containers. The upshot was that the jar's unique shape would mean slowing the production line to about half its normal speed because the bottles' shape was unstable at higher speeds. The marketer got his input from staff and plant operations just before buying proprietary molds. The marketer opted for a stock jar that would fill quickly and then invested in a labeling option to carry the "fun" message.

gives the designers and packaging developers the insights they need to translate business needs and market opportunities into a package.

PTIS Vice President Phil McKiernan puts it this way: "You need to get wide input into this document. If you use only the brand people, then you only get a brand answer. A similar thing can happen if the only guidance is on optimization and cost reduction—the package may not answer other needs. The goal is to align development with corporate objectives." Phil goes on to emphasize that input from the team we described in Chapter 4 is critical for Big P packaging to evolve. The packaging brief brings the team together to input their specific objectives; once the document is done, it also makes the objectives transparent to everyone involved in the project.

USING VISUAL IMPACT TO "SELL" CONCEPTS

The designer's talents include the ability to visualize how packaging concepts will look on the store shelf (or wherever the First Moment of Truth happens). However, many decision-makers in the value web don't have this talent, and neither do research participants. Very often, the packaging manager's job is to sell a concept to other decision-makers who may not perceive what a concept does. You may be selling to a retailer who can't visualize it on a store shelf. You may be doing consumer research where only a prototype will convey the package to respondents. Often, that function was met with "comps," detailed renderings of what the package would look like. Today, virtual renditions, mock-ups and prototypes fill that purpose and produce greater detail. They support packaging's strategic function, and packaging managers need to understand the techniques so they can use the best technology to help their target audience see what package designs can do.

Virtual design and 3D imaging

3D Simulation software is often a tool used early in the process. Software produces 3D images on display screens; they carry more information than comps. They are limited because they cannot convey the tactile "feel" of the package. However, they offer a degree of sophistication that can convey characteristics.

3D prototyping. Stereolithography and 3D prototyping techniques enhance the ability to create quick prototypes of rigid packages—plastic, metal and glass. This is particularly important as shape gains prominence versus color in visual appeal. The process uses CAD data to direct a laser-driven system that builds a full-size prototype. Some like to call the process a 3D "printer."

CAD data is the backbone of many prototyping capabilities. The data also offers another advantage. It allows use of finite element analysis (FEA). This internal computer simulation technique has roots in aerospace, and it is becoming available to work with packaging. It can analyze data on the loads a package can withstand and allow comparison of alternate configurations to see which works best. Ultimately, it could show the conditions under which a package would fail.

Samples and prototypes. Depending on the packaging materials—paperboard, plastic, metal or glass—different techniques exist to create prototypes. One place to be very careful is in creating hand samples—they must convey the finished package's perfection. Sometimes, manual assembly will create skews in alignments or less-than-professional sealing of a package. Your target audience won't accept this as a condition of a mock-up; they often see it as a flaw in the packaging.

Whatever the technique, the prototypes can reduce packaging design time and costs. They can convey emotion in a way that a rendering on paper or on a monitor just cannot do. Gauging consumer or user emotion is critical to success with packaging innovations. Many prototypes are produced for look and feel, not for functionality. Prototypes can also be very effective marketing sales tools. Prototypes can be used to optimize the offering and make the concept more user-friendly. Typically, packaging professionals will optimize size, shape, and opening forces. But marketers may want to optimize benefits, positioning support, and reason to believe. Prototypes help highlight those points.

AGING CONSUMER DRIVES UNIVERSAL DESIGN

Here's a design driver that is gaining momentum: Packaging has to meet the needs of those with diminished abilities—arthritis, reduced sight, etc. It is becoming more important as populations age in developed economies. Here's some data to show that packaging design for the aging and those with reduced abilities needs to be part of design thinking.

- In Germany, one in five persons is over 65 years of age. In Japan the number is closer to one in four.
- In the U.S, one in eight is older than 65. That is 40.4 million people.
- On average, 8,500 Baby Boomers are turning 65 every day in the United States.
- Aging isn't the only indicator of need; in the United States, 21 million adults report limitations from arthritis.

The term used around the effort to meet the needs of these people is generally "Universal Design," and good case stories already show how design accommodates those with reduced dexterity. Think of the OXO brand of kitchen gadgets that carved out a niche based on universal design. Understand how that kind of thinking can apply to packaging. Efforts gained support in 2011 when the International Standards Organization issued ISO 11156—Packaging – Accessible design – General Requirements. Couple that with work by organizations such as the Center for Universal Design at North Carolina State University, and package designers have guidelines as a basis for action. Here are some of the key elements of Universal Design:

Make information pop out. North Carolina State uses the term "perceptible information," and ISO lumps the effort under its term "characters and imagery." Doing it right may mean going beyond font, color and type size. Braille and alternative information delivery are options to consider.

Give people choices in how to open a package and make it easy. That infers an intuitive opening feature; whether it is twist or pull, the mechanism should be readily apparent. And here's one that's easy to miss—make sure it works for lefties as well as righties. ISO extends the concept to reclosure, too.

Make packages equitable to use. That means taking design to a denominator that the least able can use, but also works for Ms. Agile. It's a social issue in making sure the packaging doesn't segregate or stigmatize any users. In Japan, one example involves bottles for hospital patients; the neck is angled to allow patients lying down to use the container, yet it works just as well for upright patients.

There's a KISS principle in the guidelines, too. Make packaging simple and intuitive to use. If it takes any kind of instructions to use, it's probably too complex. Think in terms of the Second Moment of Truth. If it was hard to use the first time, you've cut down the chances of someone buying the package a second time. If it needs instructions for package or product use—can you do it with simple icons?

Know that people will make mistakes, so build in a tolerance for error. Among the things to do: Look at your communication elements and be sure they minimize hazards and errors. For things with potential error, discourage unconscious action.

Strive for low physical effort. A small-diameter cap may reduce per-package costs, but does it give people enough to grasp, and does it give them the leverage to open the container easily? Ask that kind of question for every place users interact with packages.

Think packaging end-of-life. The ISO guides also consider how a package can be disposed, recycled or reused. The guidelines suggest that the design lets those with reduced abilities accomplish those activities.

Design: A management attitude

We find the ideas of Fred Collopy intriguing; he believes that what he calls "design attitude" can extend to business management and its challenges. Collopy is Professor and Chair, Information Systems Department at Case Western University. At heart, he's a designer. From his perspective, he sees a big difference between design attitude and the usual business decision-making path.

Decision-making attitude	Design attitude
Put several options "on the table"	Put criteria "on the table"
Evaluate for the best	Explore ways to meet the criteria
Select the best	Answering the criteria produces the best solution

P&G's Clay Street Project illustrates the design attitude process; in a Clay Street Project, the first action is to immerse the team in the problem, even before looking at solutions. The more effort you can spend observing a situation's challenge, the better you can do in exploring all avenues in the integrated value web for answers. Another key to Collopy's concept lies in exploring ways to meet the criteria. At that juncture, he recommends that you observe any problem in context. The nuances you see in context (and for best results, first-hand) often carry more weight than a report that summarizes observation. Collopy, and his co-author Richard J. Bolland, explore the design attitude in their book *Managing as Designing*.

Key holistic packaging design insights

Packaging design is a system

It extends from unmet consumer needs through package end-of-life with a number of influencers along the way.

Packaging Action: Know who those influencers are for your specific market. Know their needs and values in-depth.

Design needs to get into the development process early

Too often a product development team has already specified a packaging format before bringing packaging or design into the process. That precludes some options.

Packaging Action: Aggressively lobby to get into the process sooner. Let product development managers know you can bring more options to the table earlier. You get a better understanding of product/package nuances and can produce a packaging brief that encompass all the factors that influence development.

The 'packaging brief' conveys the challenge to designers

It goes further than the traditional design brief and adds new influencers into the process.

Packaging Action: Designers and packaging developers need the entire range of needs from everyone along the value web. Be sure the documentation delivers that to design.

It can take a hands-on experience to 'sell' a concept

Once a design concept is in place, packaging managers often have to sell it to other decision-makers who probably need to actually see and touch what the concept means as a package.

Packaging Action: Know which techniques give you the best prototypes to do that.

Universal design gives the impaired 'mainstream' access to packaging

In developed markets, an older population needs help in accessing packages. Impaired persons also need help.

Packaging Action: Follow the principles of universal design to guide the packaging development process.

Make the Holistic Packaging Design approach the operating procedure in package development

Packaging Action: Use this checklist for design projects.

Holistic Packaging Design Checklist

- ❑ Consumer
- ❑ Retail and distribution channels
- ❑ Brand
- ❑ Visual
- ❑ Innovation
- ❑ Business needs
- ❑ Operations
- ❑ Universal design
- ❑ Sustainability

6 || *Know that Green is Normal*

> ***A scenario . . .*** Marketing liked the raised ink—it almost looked like a 3-D effect; it added shelf impact and a tactile sense to the package when a shopper picked it up. The ink cost more, but it was a must—in the brand manager's thinking—to compensate for the smaller front panel area on the package. The logistics manager looked at the new, smaller package as 22% more units on each pallet and how that translated into logistics savings. The new size now also matched another size in the line's packaging, generating added savings through larger purchases of containers. It used a thinner material, thanks to some structural work on the design. And the sustainability manager had a carbon footprint advantage over the former package.

Call it "green" packaging, call it "eco-friendly" packaging, or call it "sustainable" packaging. No matter what term you use, its impact on packaging has accelerated in the past decade. We've gone quickly from the search for sustainable materials to the search for sustainable systems. We've gone from small pockets of believers who worked at operational levels; now, we have C-suite occupants with the job of developing corporate-wide goals—goals that align packaging, production, and other functions.

These factors shape the way we develop sustainable packaging.

- More groups with environmental agendas exert influence within the packaging value web. Among them are retailers that want to let consumers know they share eco-friendly values, too. Non-government organizations (NGOs) also have impact. So do financial analysts who see sustainability from a dollars-and-cents view. More and more, companies believe that they need to be good corporate citizens.

- New organizations are emerging to support packaging efforts. They are resources for package manufacturers and consumer packaged goods (CPG) companies—or fast-moving consumer goods (FMCG) companies as they are called in Europe and other parts of the world.
- A cadre of consumers embrace eco-friendly values. While they are still a minority, they add pressure on retailers and CPGs.
- Packaging management has to set priorities for sustainability issues to meet corporate goals. The Triple Bottom Line—economic, environmental, and social—is no longer leading edge. In many companies it is part of the mission.

CONSUMER VALUES

In simple terms, consumer perceptions of eco-friendly packaging are changing buying decisions. The core "green" buyer is a minority, but the influence is wider than the number.

Let's look at that change by starting with the term "LOHAS consumer," a phrase that has crept into business jargon. It stands for "lifestyle of health and sustainability," and it defines a set of consumers who are concerned with the environment, social responsibility, sustainability and health for both people and for the planet. Researchers put the number of LOHAS consumers at around 20% of the U.S. market. Depending on whose definition and whose research, that number is up from somewhere around 15% since about 2005. Just as important a driver of sustainability is the added group of shoppers who base buying decisions on their perception of product or package sustainability. Some researchers call them "concerned shoppers." Added to the proportion of LOHAS consumers, concerned shoppers are about 40% of the market in the U.S. It's global, too. Concerned consumers may be a small part of the Chinese market; yet recent research suggests the number is as many as 60 million Chinese consumers. They are primarily in larger cities and would be willing to pay more for sustainable products.

A company called EcoFocus Worldwide tracks consumer attitudes on "green" issues. Linda Gilbert is the CEO, and we had the chance to talk to her at a recent conference. Here's her "take" on what the consumer is thinking. Her research identifies a "Consumer Who Cares." That's someone who takes environmental issues into consideration when shopping at least some of the time. More than 8 in 10 of U.S. consumers say they are in that category. Within that, there is a smaller group that says they are strongly committed to socially responsible choices and are willing to boycott. Packaging translates those attitudes into perceptions of your product or brand.

For "green" issues, the package defines consumer perceptions about a product and brand just as strongly as it defines any other characteristic. When these consumers buy, packaging is the gateway to the sustainability arena. Six of ten consumers EcoFocus Worldwide surveyed say it is extremely or very important to choose foods and beverages that are responsibly packaged. And 38% of respondents say they have already changed what they buy because of the amount or type of packaging. Both percentages are going up. Another two-thirds agreed with the statement, "Manufacturers need to do a better job of telling me how to recycle or dispose of their packaging."

Gilbert says that "green" issues can be a tipping point in making a buying decision. First, consumers have a base level of performance they expect from a product and package. A key factor in that base level is convenience. Packaging has to be convenient and half of the consumers in the EcoFocus research say it is a major priority. "Consumers aren't willing to 'trade down' on convenience for an eco benefit. But after those criteria are met, an advantage in 'green' or sustainable perception may tip the buying decision in favor of one brand," Gilbert goes on to explain. Sustainability can be a brand differentiator. Aligning to consumer attitudes can sometimes involve little things in packaging. Some say consumers want the recycling or certification symbols larger so they can spot them easily while they are shopping. Even a small design element like that can be a tipping point in the purchase decision.

Jonathan Asher is another packaging researcher we've come to respect over the years. He is Executive Vice President at Perception Research Services; the company specializes in packaging and shopper market research. Early in 2012, PRS shared data on a study of U.S. consumer attitudes toward packaging and the environment. The study updated 2010 research and gives a baseline against which to check attitudes. We talked with Asher about the results and got these high points.

He says that shoppers show an even greater desire to choose environmentally friendly packaging. In the recent report, 36% said they would choose environmentally friendly packaging; the number was 28% in the earlier report. However, the increase doesn't translate into good will for the companies that make "green" claims. Some 25% of the respondents said they "strongly agree" that companies overstate their packaging's environmental benefits. We suggested to Asher that the continued growth of e-everything networks feeds that belief—other research shows that consumers believe their e-everything peers more than they trust marketers.

"Most CPG companies are developing comprehensive sustainability plans that include packaging. They are reducing the amount of material they use. They're adding recycled content. They are using more renewable materials," Asher says. "However, it is important for manufacturers to educate—or at a minimum, inform—shoppers about the efforts. The CPGs have to be sensitive to the increasing degree of scrutiny shoppers are bringing to this area."

Another point from the PRS research: Consumers are getting more confused about environmental claims. In 2010, 12% of consumers said they were confused; the latest report puts the number at 20%. Our experience says that the sheer volume of environmental claims being made on packages is contributing to that. Consumers are getting more sophisticated, and they need to see clear communication to create a point of differentiation anymore. Asher comments, "If a package is recyclable, then stating that message clearly will be compelling. And if the communication can also remind shoppers to actually recycle the package, then all the better."

Sustainable packaging results from an important synergy between packager and consumer. Both want to see less material in packaging. Manufacturers save costs, and consumers claim to see it as an environmental plus. Yet, the impact is not fully visible to consumers. The PRS research asked consumers whether that has an impact on their buying decisions. About four in 10 say it makes them more interested, while slightly more than half say it has no impact. The final question PRS asked relates to buying decisions and the likelihood of an environmental claim swaying a purchase decision. According to the PRS research, almost six in 10 shoppers claim that seeing environmental claims on packages has a positive impact on their buying decision. A third claim they buy more of the brand or products they usually buy because packaging is more environmentally friendly. One-quarter claim they switch to brands or products that have more environmentally friendly packaging.

Asher observes that shoppers want to help the environment, but they need help to be effective and consistent. "While they will not compromise functionality, they are willing to pay a bit more for environmentally friendly packaging. They need to understand which packages are better for the environment, and they need to be reminded of steps they can take." Asher continues, "Eco-friendly packaging needs to deliver benefits that shoppers care about, understand, and will make use of. Ultimately, that combination reduces waste and lessens the carbon footprint."

Interesting data from the PRS research: Consumers in the Midwestern U.S. have the softest attitudes toward environmental packaging. Across a number of questions, the responses from those in the Midwest were lower than for the East, West and South. Responses showed a lower concerned about the environment and lower activity in taking steps to help save the environment. Those in the Midwest are also less likely to notice claims about environmentally friendly packaging.

Here's another piece of research. Findings from the Landor 2010 Green Brands Survey is that respondents showed an increase in environmental concerns, up an average of 3.5 percent over 2009—despite tough economic times for retailers, marketers and shoppers. Climate change was at the top of the list for consumers surveyed. Water conservation takes second place in the minds of respondents. British consumers show the most interest in reducing the amount of packaging used, ranking it second. Sustainable practice will become the cost of entry in many industries around the globe.

There's an axiom among communications professionals: Perception is reality. In this book, we're going to talk about consumer perceptions of "green" issues. We've followed those perceptions since the 1980s as part of the PTIS sustainable consulting practice. As we move through this decade, these attitudes gain some long-term significance as they become embedded in the psyche of some consumer segments. They have become reality.

How powerful can consumer sentiments be? Consider bottled water. After being a packaged product whose sales grew for decades, changing economics and consumer attitudes pulled down sales beginning in 2007. Nestlé's bottled waters include brands such as Ice Mountain, Nestlé Pure Life and Resource. The last brand is in a bottle that has 50% recycled content; other Nestlé brands have gone to extreme light weighting to address consumer concerns. Even with these efforts, industry observers say that in United States, Canada and Europe, Nestlé bottled water sales declined by 17% from 2007 to 2010. The recession drove those sales down, yet attitudes toward "responsible packaging" contributed to the decline, too. The eco attitude is also increasing the emphasis on pricing. In the face of declining market, price competition became more pronounced, impacting bottom lines.

The concern for "green" packaging has another powerful ally in the value web—the retailer. The drivers for retailers are

Teamwork gives Apple package downsizing gain

Apple gains packaging success by relying on teams of design and engineering professionals; they bring marketing, product protection and operation issues to the table in a holistic approach. One example of how it works for Apple: It downsized the iPhone package after it launched the product with the package for the iPhone. The package for the iPhone 4 is 42% smaller than the package for the original iPhone. What Apple says is important is that it gets 80% more iPhone boxes on a pallet that it did with the original container. Yes, it saves on materials, but Apple also counts its logistics savings. It nearly cut in half the number of aircraft loads to ship the same number of phones, improving Apple's carbon footprint.

Some would say Apple illustrates best practices in leveraging its work. It reports its environmental figure on its website, adopting a stance of being transparent in making information available. Among its claims: Majority of packaging made from post-consumer recycled fiberboard and biobased materials. Some find fault with Apples because it doesn't follow standard reporting formats. The Global Reporting Initiative has guidelines to encourage standard formats.

as much economic as they are concern with the environment. Consider the most prominent sustainable program—Walmart's Packaging Scorecard. It is a rating system that defined how sustainability is to be delivered to meet Walmart standards. One component is packaging "cube," the ratio of packaging volume to actual product delivered. This factor helps cut down on packaging materials by rewarding containers that have the best ratio of packaging materials to product. Yet, it also has a positive impact on Walmart's shipping costs; a key element of the Walmart low-cost operating philosophy is efficient logistics. Lower cube and higher product to package ratios helps the retailer get more product onto each truck.

INFLUENCERS IN THE VALUE WEB

Sustainable thinking is rising rapidly for both products and packaging. It has spawned a new group of industry, public and government organizations that want to shape thinking. It gets to be as much "who you know" as "what you know." Key outputs from these organizations are standards and metrics that define what "green" packaging will be in the future. It is a complex process, and the right place for packagers to be right now is in organizations that are defining sustainability. Here are some of the major organizations with close ties to packaging; the order of listing does not indicate any ranking of influence or size.

- *The International Standards Organization (ISO).* Its SC4 Packaging and Environmental sub-committee already has more than two years of work into developing a sustainability standard. Its final draft standards cover multiple areas, including source reduction, refuse and recycling. Other areas include energy and chemical recovery along with composting and biodegradation.
- *The Global Packaging Project* is a work group within the Consumer Goods Forum. It has the booklet *A Global Language for Packaging and Sustainability.* It builds the case for sustainability by probing packaging's roles and integrating it with principles of sustainability. It also examines indicators and metrics for packaging. Organizations involved in

the project have included global retailers, CPGs and FMCG companies, packaging converters, NGOs, academic and consultant. (www.theconsumergoodsforum.com/ and click on the Global Packaging Project)

- The *Sustainable Packaging Coalition* is a project of GreenBlue. Its forte is education, bringing companies up to speed on broad sustainability concepts. Its assessment software, COMPASS, helps packagers compare environmental impacts of designs using a life cycle approach. It offers environmental technology briefs and other guides. SPC gave us one of the first definitions of sustainable packaging, although several global organizations have since offered other approaches. Members include material suppliers, converters, and CPGs. (www.sustainablepackaging.org/)
- *Europen* is formally the European Organization for Packaging and the Environment. Its goal is to involve the packaging value chain in European Union and national legislation on packaging and packaging waste. It focuses on extending producer responsibility laws and package recovery; that includes the EU packaging and packaging waste directive. Members are package suppliers and FMCG companies. (www.europen.be/)
- *Ameripen* is formally the American Institute for Packaging and the Environment. In many ways it parallels the goals and practices of *Europen*. Objectives include lobbying on environmental impacts of packaging. It expects to present a perspective that includes marketing, sourcing, manufacture and distribution of packaging through its lifecycle. Members include raw material suppliers, converters, CPGs and waste recoverers. (www.ameripen.org/)
- *The Sustainability Consortium.* Its goal is to promote sustainability standards and communicate the sustainability of products. It operates in the global arena, and members span the value web. They include raw material suppliers, packaging suppliers, CPGs, retailers and others. Its board and executive group includes a number of academics. It has a packaging work group, and initial indications are that it may develop a label initiative to quickly indicate environmental attributes to consumers. (http://www.sustainabilityconsortium.org/)

- *WRAP* is an organization in the United Kingdom with a motto of "recycle more and waste less." Among its projects is the Courtauld Commitment, an effort to end growth in packaging through smarter design. Wrap says it has 45 leading retailers and FMCG companies signed up for the project. Governments in the UK fund Wrap. It also gets funding from the European union. (www.wrap.org.uk/)
- *PACNext* is a North American project of the Packaging Association. The project's goal is to help industry transition to a zero waste packaging world. Its answers include economic recovery of materials to improve reduction, recycling, reuse, up-cycling and composting. Members include retailers, CPGs, package manufacturers, raw material suppliers and others including NGOs. (www.pac.ca and click on the PACNext icon)

In addition to these organizations, a number of public interest NGOs also impact packaging. They include the Environmental Defense Fund, As You Sow, Greenpeace, and others.

A STRATEGY FOR SUSTAINABILITY

Consumer attitudes and products that add to sustainability are inputs to a sustainable packaging process. So is information on competitors' offerings. These three inputs can help you form a strategy that focuses on where you can deliver the biggest sustainable impact. Look at the "green product" resources available to you and compare them to your customers' needs. The area where you have an overlap of resources and needs creates a point of focus. Yet, the process has another component—what competitors are offering. Where your capabilities overlap those of competitors, you don't offer a unique advantage. Eliminate that sector, and you have a "Sweet Spot" where you can offer unique advantages.

With these tools, how do packaging managers integrate sustainability into their plans? Todd Bukowski is a PTIS consultant whose practice includes a focus on sustainability; we asked him about the actions to develop a strategic approach. Here are four steps he sees as being critical to effective programs:

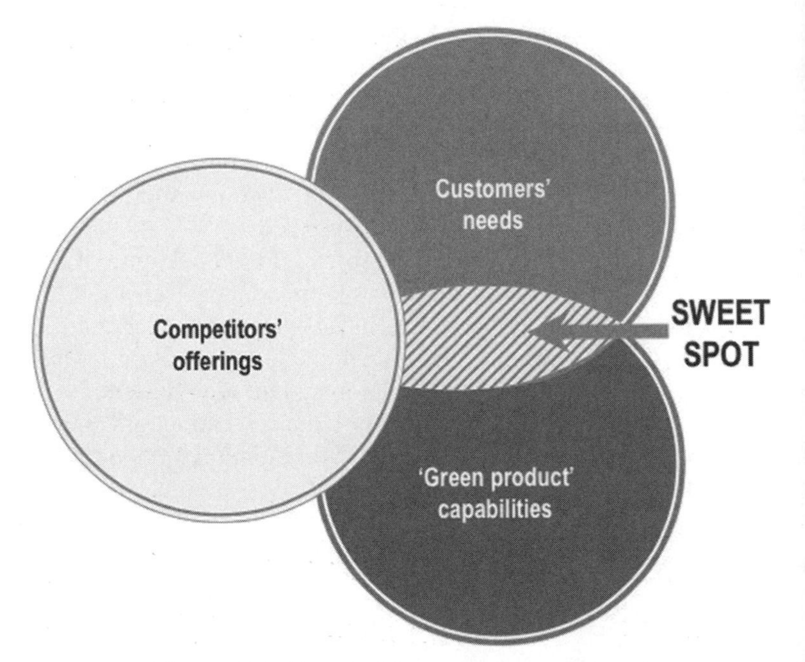

Figure 6.1. Sustainable packaging sweet spot.

1. *Understand which sustainability aspects are most important to your consumer.* We've outlined some broad consumer attitudes; yet, each company has to determine its own marketplace needs. Whole Foods, as an example, uses compostable trays for its in-store deli counters because it fits the company's market position. The Whole Foods 2012 *Green Mission Report* outlines its use of fiber trays made of bulrushes, sugar cane pulp, cornstarch, and bamboo. They replaced foamed styrene trays. We are not suggesting this as a model solution—only that it fits the Whole Foods market and the company manages packaging to meet it goals. Other retailers and CPGs may come up with different answers that fit their needs. The packaging manager's job is to understand those needs and devise tactics to meet them.
2. *Knowing how you will measure and focus efforts.* Life cycle assessment tools provide one way to measure performance. Those tools include the COMPASS product from SPC and PIQET from the Sustainable Packaging Alliance in Austra-

lia. (www.sustainablepack.org/default.aspx). They quantify how different packaging solutions compare. Standards such as the Walmart scorecard and supplier measures such as those from P&G and Unilever also offer ways to measure performance. One caveat is citing corporate scorecards—they are weighted to meet the goals of the company and may not reflect a broader consensus of how to value components. (See the analysis below on the Walmart scorecard.)

3. *Develop a materials strategy.* Materials are not inherently sustainable. They are sustainable in a use context. Consider the process of looking at a material's pros and cons based on the use conditions. A materials strategy is going to be specific for each specific company depending on its primary packaging format, materials, distribution channels and market. It will also depend on the number of materials in packages; for example, the adhesives in a gluing operation become part of the strategy. No matter what your materials, once you have a list, here are criteria by which to assess the materials.

    - Environmental impact. You will need to work through a range of influencers. They include those at packaging's end-of-life—landfill operators and recyclers. Retailers are important, too; they drive material choices with their policies toward materials such as PLA, PET and others. Watch for attitudes on issues such as compostability. Organizations such as SPC may be the best way to navigate the environmental landscape.
    - Cost. If you pare your materials list, you may be able to leverage your cost influence as you buy more or fewer materials.
    - Recycled content. One important metric is the percentage of post-consumer recycled material in the container.
    - Renewably sourced. Bio-based materials support this criteria. Also, be sure you can identify a chain of certifications that validate the sourcing to the original point of growth or harvest. One example is sustainable forestry where certifications appear on finished packages. Converters and packagers need to be sure fiberboard suppliers can validate the certifications they claim.

- Carbon footprint. Look at lifecycle assessment programs to assess and reduce your materials footprint. Typically, it's not necessary to pay for a formal LCA. They can be costly. You can get direction on your package design with simpler tools such as COMPASS and PIQET. The more formal LCA studies have a role in identifying big areas in your sourcing and operations where you can optimize environmental issues.
- Optimized SKU and distribution efficiency. A component of carbon footprint can include energy used in transportation. A package's physical configuration contributes to transportation efficiencies. Cube optimization programs such as CAPE and TOPS aid in reducing cube and getting more product on a truck, in an airplane, or on a railcar. Here's where to use Holistic Packaging Design thinking; some say that 80% of the carbon footprint is determined at the design stage.
- Safety and quality. We talk about risk management later in this book. A materials strategy is an integral component of that. Consider bisphenol A (BPA). The FDA sees no reason to ban it—for now. Yet, NGOs pressure food packagers to eliminate it. For nanotechnology, momentum has built to exercise caution in adopting it, especially for food packaging. Evaluate the materials that are important in your packaging process to see which could present safety risks.
- Laws and regulations. The devil is in the details because regulations can impact minor material components. For examples, adhesives can be affected. Europe's REACH program has a global impact and you need to understand it. Definitions such as "degradable" and "compostable" will emerge from laws and regulations. The Federal Trade Commission continues to work on its "Green Guides" update.
- Sourcing. That includes both global and local sourcing. Local country sourcing may address the social element of sustainability, and be ready to track your sourcing in terms of your company's corporate social responsibility goals. Consider third-party certification programs for your sustainable sourcing programs.

4. *Communicate.* Finally, packaging needs to communicate efforts to stakeholders in their company and along the value

web. Among targets: Vendors, so they know your sustainability practices and goals. Devise ways to engage vendors in the process. And don't forget employees. Can you make sustainability a part of each employee's annual review? This usually resonates well with employees by recognizing their inherent willingness to support sustainable practices. Here's an example of one company's process in developing its message: When Stonyfield Farms switched to PLA yogurt cups from polystyrene, it did a lifecycle assessment to compare new and old containers. It showed the new container created 48% less greenhouse gasses. Yet, the new material wasn't perfect. Like the earlier polystyrene cups, PLA could not be recycled. However, better greenhouse gas numbers, reduced breakage and better potential end-of-life scenarios tipped the scale in favor of the change. Stonyfield Farms put all those decision points into communications aimed at stakeholders including consumers, customers and the media.

METRICS AND THE TRIPLE BOTTOM LINE

Developing metrics is an on-going process that involves the entire value web. Those metrics point to where we, as a society, go with sustainable, eco-friendly and green packaging. Look at the value web we outlined in Chapter 4; then look at the influencers outlined above. The effort is collaborative effort. Not without disagreements, and not without winners and losers, but an effort where everyone along the value web can participate. We believe that by participating in the process, companies and organizations can influence the impact it has on them.

Here's the big change the process has delivered: Before sustainability, the dominant metric was cost. In particular, it was cost within an organization's own accounting system. What sustainability is doing is adding social costs such as environmental degradation. It is reallocating disposal costs, and it is accommodating people costs. That's why sustainability is a three-legged stool—economics, environment and people. The biggest changes in the past decade have been incorporating non-economic costs into the packaging equation.

The emerging metrics of sustainability set forth the path packaging has to take. Knowing that path along with your objectives can help you develop the priorities that are appropriate for your operations. Among the tools available are lifecycle assessments and scorecards such as those from Walmart. You need to understand how each of these guidelines influences your packaging choices. We've included this caveat before: Metrics like the Walmart scorecard focus on the developer's goals and may not reflect a wider, community approach to measurements.

Lifecycle assessments. There are a number of them available to packagers, and they primarily address the environmental side of packaging. COMPASS on line software from the Sustainable Packaging Coalition is one example of the process. It incorporates several factors:

- Consumption metrics include fossil fuel, water, minerals and biotic resources.
- Emission metrics measure greenhouse gases, human impacts of clean production and aquatic toxicity.
- Packaging attributes include recycled content, source-certified materials, solid waste and material health.
- Life cycle phases capture information on the impact of material manufacture, conversion, distribution and end of life.

Scorecards. These programs are metrics, and in devising algorithms for these tools, the developers make some judgments about values of various factors. They may be government, NGOs or businesses. Let's use the Walmart Packaging Scorecard as an example. It is not the only approach to assigning value to sustainable factors. P&G and Unilever have developed similar approaches for their suppliers. We will use the Walmart system because it has enough experience behind it to offer an extended look at its impact.

One Walmart scorecard component is packaging volume versus product volume. It can lead to use of less material per unit of product. It also helps another of the components called "packaging cube"—the volume defined by the package's outer dimensions. Those two components impact other environmental factors along the value web, particularly in the distribution cycle. A lower cube for a given amount of product can mean more products per pallet. That can translate into more product per truckload, a significant factor for Walmart (a company that some say is among the country's top logistics practitioners). High fuel costs make the process ever more important.

The Walmart scorecard also influences packaging procurement by assessing a value to the distance to transport packaging components. The incentive is for in-house or local procurement of packaging components versus national or international sourcing. The scorecard also has a human component in the material health and safety section; it looks at the toxicology impact of manufacture on communities and ecosystems. It also includes OSHA injury rates associated with packaging manufacture. Another feature of the Walmart system is that it "grades on the curve." Scorecard results for any given package are not compared to a standard, but rather than to other packages in the category. That way, Walmart set a ranking on packaging that is just one factor in the overall Walmart buying decision.

Understanding those factors is input at the packaging design stage where fundamental decisions can significantly affect results.

PACKAGING, END-OF-LIFE AND EPR

Among green packaging elements, the concepts of the package's end of life and the regulatory model of extended producer responsibility (EPR) are among the more recent eco-related dimensions. Both are attempts to add metrics to actions that might have simply have been "good for the environment" only a few years ago. Recycling and reuse of packaging probably go back to the first wine casks. Yet, today's "disposable" societies change the emphasis—for good reasons, very few packages can be refilled today. Disposal in a landfill was a low-cost option until rising dollar and social costs grew into obstacles. Now, the emphasis is on end-of-life somewhere other than in a dump. And making sure that happens is the function of the package-producing levels within the value web.

The first option in dealing with end-of-life and EPR is to optimize the amount of material in the package in the first place. It is consistent with lifecycle assessments and scorecard directions.

It's not a new idea. Packaging schools have been teaching the concept's basics for a half-century and more. It was taught in terms of a systems cost impact rather than environmental impact, and the principles are parallel. Here's a simple example: If you drive a truckload of loose tomatoes from farm to a city grocery store, costs include damaged tomatoes and the clean up. Put the tomatoes in baskets and you provide some protection, packaging cost goes up and losses go down. Put six tomatoes in a blister pack—loss drops further and packaging costs rise. Somewhere there is an optimum point where the total cost is lowest; adding more packaging beyond that point costs more than the total system savings. In that scenario, disposal of the baskets or the blister packs weren't the packager's responsibility—refuse collection took care of it.

In an analysis of packaging consumption, Europen makes a similar analysis of package to product ratio, and underscores the point that the ratio has to be compared against social trends. It particularly notes that for smaller households in developed societies a higher package to product ratio may make sense as it reduces total environmental impact. That could be particularly true for perishable items. The emphasis on sustainability and eco-friendly packaging changes emphasizes a holistic approach that

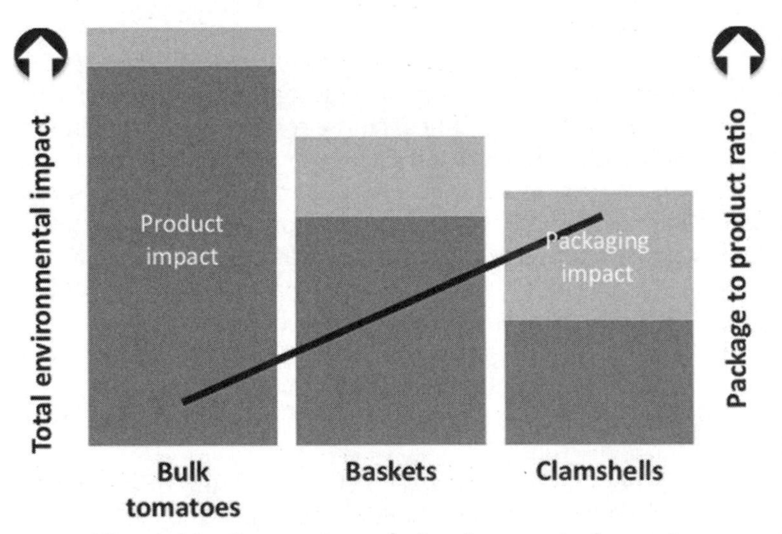

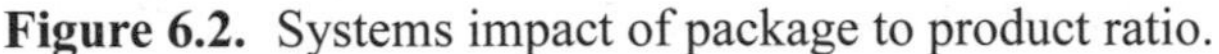
Figure 6.2. Systems impact of package to product ratio.

considers all elements rather than focusing on a narrow metric. What approaches such as scorecards and EPR regulations do is give us ways to quantify the environmental impact in a way that can't be done from an economic model alone.

Extended Producer Responsibility laws

The major driver to focus on end-of-life and disposal of both products and packaging are the Extended Producer Responsibility (EPR) laws in Europe. They form a model that is being translated into a form that works in Canada. The expectation is that similar guidelines will come to the United States. In both North American countries, EPR laws and regulations are emerging on a provincial or state level rather than being coordinated by federal governments.

In essence, the laws assign responsibility for end-of-life events to the product or package producer. The European model relies on a third-party administrator to handle the actions. They are commonly called producer responsibility organizations (PROs), and generally work within a specific country. In Europe there is an "umbrella" called PRO Europe; its official name is Packaging and Recovery Organisation Europe. Its mission is to support individual national organizations.

One of the reasons sustainability has grown so rapidly is that it can be a win-win situation for everyone along the value web. Economic impacts continue to be high on the lists of priorities for raw material suppliers, package converters and packagers. However, they usually compliment the environmental and human concerns of the retailer and consumer. The concept of innovation has been successful so that the majority of solutions deliver economic and environmental benefits.

Green Science design as an immediate answer

Sustainable answers that deliver economic and environmental benefits are usually based on materials technology points. There are answers emerging for the future, but we believe there are valid answers for today. They lie in the concept of "green science design." It builds on a holistic technology assessment process that delivers the cost-savings that manufacturers need, the sustainability that consumers demand and addresses the big-picture drivers. Here's what puts green science design on the top of the list: It is both doable and available. Here are the foundations on which green science design is built:

1. Focus on preventing waste, not on disposing of it later. For example, the electronics industry has already moved manuals for CDs and other "hard" media to Internet downloads, reducing materials use. The Sonoco Sustainability Solutions (S3) program, as an example, audits facilities and reviews waste management efforts. Audits like these can point to areas where materials can be eliminated.
2. Employ processes that avoid auxiliary materials and by-products. A corollary is to use catalysts rather than consumable co-process agents.
3. Use materials that are safe and non-harmful. Even if long-term experiments show BPA to be safe, its image as a "bad actor" shows the damage that could result from such materials. Plant-based alternatives to BPA are already under development. Keep them in your crosshairs as you do your technology scouting. One example is the S.C. Johnson Greenlist process; it classified ingredients by impact on the environment and human health.
4. Energy efficiencies are part of the design.
5. Renewable feed stocks. While long-term research will give us more bio-based materials, there are already niches where those materials can be used. Plant-based PET already has a niche in beverage and food containers. Molded fiber food trays have a place in foodservice; in particular note the use of fibers from sources other than trees—bamboo, bagasse/sugar cane, and wheat fiber to name a few materials.

   Action: Keep your technology scouting focused on renewable feed stocks for the materials you use.
6. Design for end of life. One target is degradation. The Frito-Lay SunChips degradable chip bag stumbled, but that is something to be expected when a packager goes out on the leading edge of green science design. Even though he bag material's initial design was just "too noisy" for consumers, Frito-Lay pushed for a degradable solution. And they found

it. It may be a classic study in green science design with an initial solution, a setback, and a new solution based on persistent technical investigation. In 2011 Tetra Pak launched the Tetra Evero Aseptic bottle in Europe and South America; its plastic top separates from the paperboard body as a way to get materials into the proper waste stream.

The approach has metrics in life cycle analyses and life cycle assessments.

Key insights on green packaging

Consumer attitudes alter packaging goals

The percentage of people who base purchase decisions on "green" packaging may be as high as 40%. Yet, consumers won't trade off convenience or functionality for green benefits.

Packaging Impact: Know where consumers of your product category stand on environmental issues and what they see as functional and convenience needs.

New influencers along the value web are changing packaging priorities

NGOs and industry associations may be the most important groups defining sustainable, green or eco-friendly packaging goals.

Packaging Impact: Know what organizations may impact you and where to find help.

Develop a materials strategy to support green packaging

Look at what you are using against a grid of environmental characteristics.

Packaging Impact: Ask questions such as: "What's the recycled content?" "What's the carbon footprint?" and "How can I optimize distribution efficiencies with packaging?" Communication with stakeholders becomes an important part of the development process.

Metrics embrace economics, environment and people

This is the "triple bottom line" that sustainability brings to business.

Packaging Impact: Know how deeply your management incorporates each of these principles in its corporate objectives. It could mean that a sourcing criteria might be a supplier's performance record against OSHA regulations.

Packaging end-of-life issues have risen in importance

Assuming a package will go to a landfill is no longer a good assumption.

Packaging Impact: Put end-of-life issues at the earliest phase of the packaging development process. It can be solved best at the design stage.

7 || *Global Growth is for the Taking, with Caution*

> ***A scenario . . . Nairobi, Kenya.*** The cabbage is plentiful in the supermarket, along with other local vegetables. What brought them there was a non-government organization that worked to reintroduce native crops into local agriculture; with their ability to withstand climate swings, native crops bring back some food security lost when Western crops dominate. The crops also fit local farmers who tend small plots. After the NGO got farmers back into native crops, it organized them into a producers' co-op that could reliably deliver quality produce into a distribution channel that needs quality and uniformity. A multi-national company worked with the NGO to develop a gas-flushed bag that fit into local produce boxes; the system had to be inexpensive and simple for an unsophisticated environment. It added shelf life that kept produce fresh. It also reduced food waste in a part of the world where the loss from farm to consumer is critical.

Any examination of packaging's global role needs to start with a look at global macroeconomics; they define where the opportunities lie and their nature. A logical starting point is China, where the economic growth rate in first quarter 2012 was at an annual level of 8.1%. Yet, those numbers also suggest some uncertainties. Accompanying the data were headlines saying, "China's economic growth falls to near 3-year low." And while China may make the headlines, the trend of slowed growth extends to other Asian economies and India, where growth rates of 10 to 12 percent were common. More recent projections put growth in a 6 to 8 percent range. These shifts may be an indicator of some basic global changes. Analysts suggest that slowing growth results from lower exports from these countries. Some economists say the change is necessary for global economic stability.

Here's what the International Monetary Fund (IMF) says about the change. Its 2011 World Economic Outlook says: "Advanced economies with current account deficits, most notably the United States, need to compensate for low domestic demand through an increase in foreign demand. This implies a symmetric shift away from foreign demand toward domestic demand in emerging market economies with current account surpluses, most notably China." The IMF report, dated September 2011, goes on to say, "This rebalancing act is not taking place." IMF data in mid-2012 confirms that earlier caution, suggesting that China's growth rate would be 8% in 2012 and 8.5% in 2013.

Echoing the belief that the change is not just an issue for China, the IMF report also notes that other economies in Asia are running surpluses and they are looking at building domestic demand. The term "BRIC countries"—Brazil, Russia, India and China, has crept into the global trade jargon; as a group, it too has to be included in any look at global economics. One economist explains that this group of countries has grown to the point where they exert real influence on the global economy. The group's strength emerged in 2008 when BRIC countries were able to adopt their own policy responses in what was a huge economic crisis; that kind of response was beyond their resources in the past. The pace and direction of global change is also underscored by the recent strength shown in Sub-Saharan Africa. Some use the term "pole of global growth" in describing the region's economic growth; projections call for growth in the region to be as high as 6 percent for 2012. Figure 7.1 represents some long-term global projections out to 2035; they show that even for an extended horizon, higher-than-average growth lies in developing economies.

Another term creeping into global economic jargon is CIVET—Columbia, Indonesia, Vietnam, Egypt and Turkey. These are economies with young populations that are seeing rapid growth in disposable income. Accompanying that is potential growth in consumer goods—cosmetics, beverages and foods in particular. These are economies where smaller-sized packaging respond to growing, but lower, consumer incomes.

Beyond those numbers, some economists see a global rebalancing already underway. We don't know if it is happening as fast as some economists—like those at the IMF—would like to

Region	Per Cap Income 2010[a]	Annual GDP Growth 2008–2035 %	Pop. 2015 (000s)	Pop. 2020 (000s)
Africa, Sub Saharan	1,963	3.7[b]	963,752	1,084,318
Brazil	9,390	4.6	202,826	209,575
China	4,260	5.7	1,364,978	1,381,592
Germany	43,330	1.8[c]	80,583	79,720
India	1,340	5.5	1,249,867	1,324,347
Russia	9,910	2.6	140,754	139,306
United States	47,140	2.6	321,698	334,635

Source: World Bank, U.S. Energy Information Administration.
[a]US Dollars adjusted for purchasing power parity.
[b]Total projected growth for Africa.
[c]Total projected growth to Euro Area.

Figure 7.1. Representative economic growth.

see it. However, indicators say it is having an impact on developed countries and the way they market and package goods. For example, a Goldman Sachs economist suggests that Germany is setting a lead with its exports to China; in 2012 they may exceed exports to its neighbor, France. In the United States, anecdotal evidence is mounting to show production moving back to the U.S. from China. That scenario shortens the supply chain, reduces lead times and may add some pricing stability; when a systems analysis is applied, the advantages of offshore production aren't as big as they may seem. One category involved early appears to be small electronics, and the media has already coined the term "re-shoring." With the move comes packaging business in the United States—raw materials, converting and the equipment to package products.

We talked about the global numbers and packaging trends with John Mahaffie. He is a principal of Leading Futurists, LLC, and we've worked with him over the years on a number of key research projects. His focus includes packaging, and he is able to blend the big-picture numbers with the capabilities of the industry. Our discussion with John suggests the macroeconomic trends may impact packaged goods and packaging this way:

- The emphasis on technology will increase, and the technol-

ogy scouting activities we detailed in Chapter 4 will gain even more importance. Here's the reason: Mature economies simply will not see labor costs anywhere as low as those in the developing economies. However, application of systems thinking, holistic approaches, and total cost of ownership concepts will show that countries with higher labor costs can be competitive. Technology, however, is a double-edged sword. For developing economies, it may provide a way to leapfrog into higher productivity and quality that increases the ability to compete in the world market.
- Businesses looking to grow through export have to learn new ways of developing products and packaging for new markets. It is far more than learning to speak the local language and follow local customs. Businesses that want to grow in emerging markets need to learn, in depth, about consumers is in emerging markets. They need to learn intricacies of distribution channels and how to use design to meet local needs. The global value web offers the power to do those things, and using it effectively is one step toward success.

As we talked through this macro scenario, our discussion with Mahaffie pointed to more trends that may evolve and impact raw material suppliers, converters, and retailers. The impact will also fall on consumer packaged goods companies (CPGs), as they are known in the United States, or fast-moving consumer goods companies (FMCGs), as they are known in much of the world.

- Enormous growth for packaging. It comes from increasing buying power and from the millions of people who join the package-using economy each year. When looking at the growth, look particularly at the "bottom of the pyramid."
- Emerging markets are not necessarily built on old technology. Sometimes they will quickly advance packaging forms and practices. Certainly, the Chinese printing industry is a prime example of that and the move to leading edge printing equipment helped China compete by offering high-level graphics on packaging for export. The growth of cell phones in Sub-Saharan Africa may be a driver of economic advances in that region. The export of aseptic dairy packaging

to developing countries represents state-of-the art processing that address a critical element in those countries—reducing food waste from farm to consumer and delivering social benefits of improved nutrition and health.

- Developing markets spur innovation. The packaging development process will have different inputs than in developed markets. Consumers are different. Distribution channels are different. If product and package developers apply the holistic process we've stressed in this book, the results are going to more often be innovative rather than incremental.
- The developing economies are not only the next opportunity. They are the next competitors. Local entrepreneurs in the emerging markets are making strong plays in their own domestic markets. Check the Internet for companies such as Cavin Care in India and Wahaha with Future Cola in China; both firms tout their achievements in growing their domestic markets. FMCG companies like those can establish strong local brands and they can develop their own export capacity. They offer competition for the packaging and packaged goods industries in developed countries.

Let's examine the potential by looking at a segment of emerging markets that simply doesn't have a parallel in developed economies. That is the "bottom of the pyramid," or BoP for short. It encompasses 75% of the world's population. Figure 7.2 puts the global wealth distribution in perspective. To give the global income numbers meaning in context, they are adjusted country-by-country in terms of "purchasing parity power," a factor that equalized the purchasing power in each economy. For BoP consumers, top earners make less than US$ 10,000 per year, and most make less than US$ 2,500 per year.

A caution in looking at the group that earns US$ 2,500 to $US 10,000 per year. It is the middle range, but it does not always equate to the middle class in developed countries. Often, they are distinctly different with less wealth than those in developed economies. Yet, they do reflect a Western-style middle class with aspirations that parallel to those in developed economies. They are moving to cities and suburbs; they are on line and getting mobile phones. They are buying branded packaged goods. For marketers who think in terms of the U.S. middle class, think

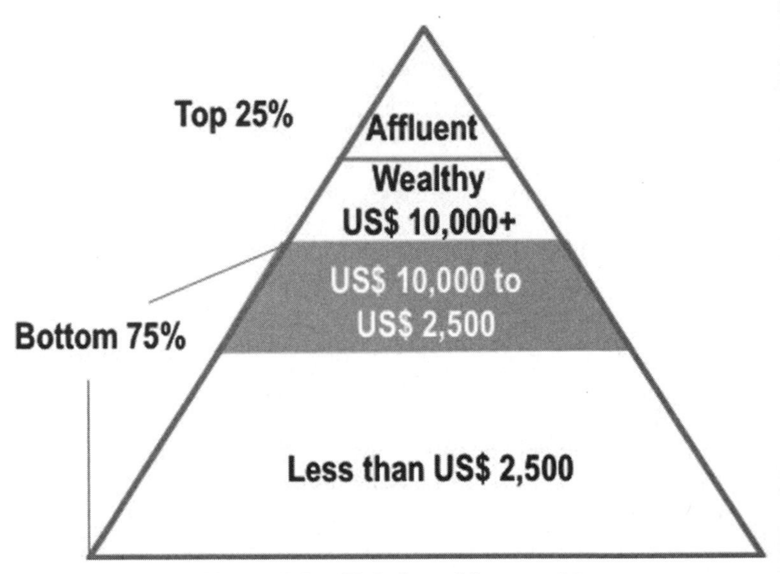

Figure 7.2. Global wealth pyramid.

more in terms of that group as it was in the '50s, '60s or '70s rather than where it is today. Expect the global middle group to grow as a proportion of the population. A recent report from the Brookings Institute focusing on Latin America says that, by 2030, half of Brazil's population will be middle class. The report expects the number to be 80% in Mexico.

- Look at Sub-Saharan Africa and South Asia as examples of BoP markets where incomes are under US$ 2,500 per year. Packaging capabilities in these economies are basic, with some suppliers who produce commercial quality containers and some who produce export-quality packaging. Huge markets with the tightest of margins.
- In the mid-range on the pyramid are emerging markets where annual incomes approach $US 10,000 on a purchasing parity basis. Packaging is gaining in quality and variety; international brands are appearing. These economies need faster operating cycles and design and technical skills. See Asia, Latin America and the Caribbean in this group. Eastern Europe is transitioning out of this group toward the advanced economies.

- Advanced economies represent the top 25% of global markets. They are where innovation, advanced technologies and customization are the norm. The U.S., Germany, Japan and South Korea are examples of these economies.

DIFFERENT REGIONS, DIFFERENT STRATEGIES

The way the bottom, middle and top of the pyramid respond to economic growth differ, depending on the region. Here are some key characteristics across major regions.

East Asia and Pacific. The region's biggest segment is bottom of the pyramid—hundreds of millions of consumers who are just entering the packaged goods society. A top priority is to advance food packaging for this segment. A focus will be on local markets and domestic consumption; one example is Chinese beverage maker Hangshou Wahaha. It built a US$ 5.2 billion business by serving rural areas. Overall, top issues with an impact on packaging include quality, safety and anti-counterfeiting. In this region, social issues will help shape business strategies; issues will include workers' rights. Rising labor costs and intellectual property theft are potential risks.

Euro Area. This is a "retired" continent, yet the dominant theme is the "new" consumer who looks for experiential connections with products and packaging. The debt crisis and low-growth economies shape business responses to the market. In addition to the debt crisis, Europe also presents risks of increased eco-demands.

Eastern Europe and Russia. This area is moving into the middle range of economic development; in some ways, it shows a "split" personality. More areas see the growth of modern retailing; that is driving the continued growth for branded, packaged goods. It also reflects some characteristics of the "new" consumer including those who embrace "green" values. Yet, pockets of traditional retailing remain, and Russia's political environment and corruption create obstacles for growth. This region does not yet have the ability to set its own economic future, and risks include the region's vulnerability to global economic cycles.

Latin America. This area is moving toward a strong regional marketplace; some say there is a de facto Trans-Andes common market emerging. The region sees retailing and consumer packaging modernizing quickly. The emphasis on sustainability issues are surprising, especially in Brazil. If economic growth relies on political stability, than Latin America presents risks from political instability. Corruption and intellectual property theft are other inhibitors to growth.

South Asia. The key segment is bottom of the pyramid, and the focus is on food packaging. The domestic market drives growth, yet it remains connected to traditional retailing. Its product channels are singular in a reliance on micro-distributors to reach bottom-of-pyramid consumers. Regulators will work with that system rather than quash it. Yet, Western-style retail is also emerging here. Expect the rise of "modern" India to shape product and packaging responses. Political instability presents risks in this region.

Sub-Saharan Africa. Deep poverty continues; with that comes immediate bottom-of-the-pyramid opportunity. Packaging growth will come as people move into a packaged-goods, cash society. Notes one observer in Kenya: "The reality is that the middle-class now shops from the supermarket once a month, but goes to the kiosk more often to replenish." The Chinese lead the pack of foreign developers coming into the market; it is not a focus on packaging, they are developing infrastructure, housing and working in a number of areas. African companies are emerging with the capability to dominate local markets and compete globally; we're going to see African brands in the global marketplace. One regional characteristic that affects growth potential: Nearly everyone pays and banks by phone.

North America. This is the market-to-me society, and it is moving toward new consumer values. Some refer to the "new consumer" in recognizing the scenario; other uses the term "post consumer" society. Use what works for you, yet know that psychological desires often determine its needs; the term experiential packaging is a response to that scenario. The "new" consumer says, "customize for me, give me experiences and novelty, and I'm going to use social networks to share what I think." It's too early to tell if "re-shoring" offers significant domestic growth, and there is risk in the ability of foreign competitors to "leapfrog" technology and be a significant factor, even in an economy trying to bring production back to the region. The ability of businesses in the region to leverage technology to compete with lower labor costs of other regions is a key.

TARGETING GLOBAL MARKETS

If you are going to compete in emerging and developing economies, the questions to ask are: Where do you want to play? Can you drive profitable sales? When you answer those questions and have a direction, then the exercise becomes a strong parallel to the path of knowing consumers and the distribution channels we outlined in Chapters 2 and 3. You have to know those characteristics in depth for specific countries and markets. C.K. Prahalad was a global business consultant who put it this way, "Focus on immersion with the new consumer. Learn about their needs and priorities." It is the backbone of success in developed markets, and it parallels the strategy for success in current markets.

These insights underscore the paths outlined in Chapter 2 on knowing the consumer. The paths in Chapter 4 on the Integrated Value Chain also apply to emerging markets. In particular, know that you have to include non-government organizations that shape economic developments.

Mahaffie says his experience points to some general guidelines toward developing products and packaging for the BoP, emerging and developing markets:

- Target low-price affordability. Low price is important, but consumers in these economies also want quality, convenience and many of the product/package traits we see in developed countries.
- Find ways to deliver products that consumers thought were not available in their rural areas or were beyond their economic reach. Sachets are a common packaging format, and kiosks are a common distribution channel.
- Compete on value, quality and brand. And don't forget lifestyle, cultural and aspirational fits. Brands will often be new, and the emerging market shift overall is from loose to branded, packaged products.
- Think in terms of "frugal innovation." Packaging develop-

ment and packaging costs have to be held to a minimum. R&D focuses on low-cost packaging.

- Be ready for rapidly changing business conditions. Urbanization may be rapid, and retailing may advance quickly. Technology and distribution channels may leapfrog the patterns we've seen in developed economies.
- Corporate social responsibility has to be part of your business DNA. Government and non-government organizations often drive development in emerging and developing economies; their agendas embrace worker rights and environment. Your "competition" may be not-for-profits and "social enterprises." They may also be your collaborators. The emphasis on social goals and agendas should raise caution, too, when perceptions put marketing efforts in a bad light. One global marketer employed a riverboat on the Amazon River in Brazil; its tactic was to reach communities deep in the region. The tactic earned negative headlines with the gist of "selling junk foods to Brazil's slums." In emerging markets, governments and NGOs may be bigger players in the integrated value web than they are in developed countries.
- Local sourcing may be required. This is connected with the social enterprise focus of governments and NGOs in emerging markets. Local sourcing will often require investment, training, etc. to help local partners reach adequate quality and production levels.
- Local competition can emerge quickly in fast-evolving markets. And look to nearby markets for competition, too. Those local businesses may see adjacent markets as a path for their expansion, too.

The last mile in distribution

The last mile of distribution in BoP and emerging markets is markedly different than in developed economies. It demands attention often not needed in the more advanced markets—both from a marketing strategy and product protection perspective.

From the marketing perspective, a prime need is often to develop a trusted cadre of small distributors. Often they sell out of kiosks. Product marketers may need to finance this network and make each member a micro-franchisee. Network members—of-

Indian shampoo wins at bottom of the pyramid

In India, Chik shampoo is a classic case story that shows how packaging and marketing to the bottom of the pyramid can win. The goal was to bring this personal care product to rural and semi-urban markets, and the parent company, CavinKare, did it before bottom-of-the-pyramid was even on anyone's radar. The product had to be affordable to women aged 16 and older; communicating its benefits and just getting it to people were challenges faced by what is today one of India's oldest brands.

The answer to affordability was a sachet, an innovation in the Indian personal care industry at the time of its launch. The sachet's film used the graphic of a woman with flowing hair to convey its primary selling point—a quality formula that delivered results. It's low cost fit the BoP market, and it easily fit into a system of micro-distributors, largely in Southern India. An advertising strategy leveraged cinema stars in a region where cinema was a central means of entertainment; it was augmented by radio.

The Chik brand is from CavinKare and has held as much as a 35% share of all shampoos in rural Indian markets. The company manufactures and packages in one facility and has contract packagers produce some of its product. The company exports to Nepal, Sri Lanka, Malaysia and other markets.

ten women—need to become collaborators. They help explain the local system to manufacturers and find gaps that can be used to advantage. They may not have the skills to do it well. Training is one answer, yet remember packaging's role as the silent salesperson. Packaging's role at the First Moment of Truth is every bit as important in developed economies, even if the FMOT is in a kiosk. The e-everything revolution has a critical role here with mobile phones a key to that collaboration.

From a technical packaging perspective, the last mile may present the worst-case scenario for product protection. The CPG has less and less control over uncertain transport and storage conditions and the package gets to the ultimate consumer. Protection requirements need to fully accommodate those conditions.

In emerging markets, compared to BoP, retail is moving from disorganized to organized. Said another way, it is moving toward a developed market system. In India, for example, the entire retail industry was about US$ 350 billion in 2011; to put that in perspective, Walmart posted US$ 260 billion in sales for its worldwide operations. However, Indian retail is expected to grow at a rate of 5% per year into the next five years.

MARKET ENTRY STRATEGIES

We also talked with Mahaffie about where to get into the market. He suggests segmenting the market by consumer buying power—low, middle and high. He also offers this caveat in that strategy: middle class in emerging markets does not equate to middle class in developed markets.

Going in low

Going in low means to create formats that target the poor and to develop product, package and distribution tactics that fit local situations. If you follow the axiom that the consumer is first, that means packaged products that improve low-income households' access to health, food and nutrition. It is one of the reasons food packaging is seen as a priority. Marketing has to include word of mouth, demonstrations and free samples to show new consumers the product's benefit. The consumer defines quality. That statement is as true for consumers in developing economies as it is for the "new consumer" in developed economies. Advances in communications and the ubiquitous mobile phone will create the same kind of consumer-defined quality as social networks do in developed economies.

Going in low also positions a brand for growth. Global economic data all suggest a rising level of economic participation for the lowest ends of the market. By developing a brand that fits those at the lower ends of economic classes, the brand will rise as their incomes rise. For packaging, that can mean redesigns to maintain brand loyalty while the value proposition evolves for higher-income consumers. Packaging formats may change, too; a brand may start as a sachet, but growth will involve more sophisticated formats. Attention needs to be paid to changing distribution channels. In India for example, the growth of Western-style retailing adds packaging demands.

Going in higher

Those consumers in developing markets who are moving to and beyond the US$ 10,000 annual income level respond to business models used in developed economies. They will need to be adapted to local conditions, but several components work. One is to offer variety and luxury, but do it in a way that is a cultural fit; packaging design and marketing are key elements here. Prestige appeal works here. You have to develop consumer insights, and leverage technology; it can put a multi-national at an advantage to a local business that might not have the needed expertise. In Brazil, Kraft's found an on-the-go market segment for its Club Social crackers; Kraft sold in small packages that fit in backpacks and pockets and saw sales rise.

Go in high

Consumers who are affluent often want modern, Western brands and luxuries. In China, for example, if that segment is less than 5% of the market, it is still 65 million consumers. Think in terms of the right side of the well curve we detailed in Chapter 3. The product/package need to be on the premium side of the curve; your distribution channels also need to be on the right side of the curve with prestige and "boutique" attraction. Think luxury rather than value, and use experiential packaging as a strategy.

Market development strategies

Tetra Pak is betting that rising incomes of bottom of the pyramid consumers will mean growth in the milk and beverage industries. Its target is economies such as Nigeria and Kenya, and one driver is the outreach to low-income consumers. In fact, Tetra Pak goes beyond the typical "bottom of the pyramid" approach and says it wants to reach those "deep in the pyramid". This is also a case where the technology fits a market place need. Tetra Pak's core technology is aseptic packaging for milk and beverage products; as a result, the packaged products don't need refrigeration to get through the distribution chain. That's important in emerging economies; there the so-called "cold chain" link of refrigerated distribution is virtually nonexistent. Tetra Pack is pinning its hopes on an estimated growth of milk consumption in Africa from about 15 billion liters in 2010 to about 25 billion liters in 2020. The effort isn't without challenges. One is alternative milk packaging in plastic bags; it is an affordable alternative in the local market. Another is the environmental issue. Even at the bottom of the pyramid, recycling has the attention of both governments and non-government organizations.

Unilever is another multinational that sees growth in developing markets. Estimates say just over half of Unilever's business is in emerging markets, and it expects that proportion to rise to about 75% by 2020. Each day, two billion people buy a Unilever product. Market observers say one Unilever tactic for emerging markets is to look at high quality at lower prices. That position can be a plus for global brands, in part because local brands may not be as high in quality. Data developed in a recent report suggest that Unilever saw its volume grow by more than 10% in

Packaging grows faster in developing markets

What's the size of the global packaging market and key countries within that index? Depends on who you ask, and we've put together a consensus list based on a number of research reports and the authors' insights into the markets. What is worth taking from the list is the trend-line for growth. It all lies in developing countries. The growth for the U.S. and German markets reflect developed countries with 1.3%. That's just less than half the global growth rate and well under what is seen in the BRIC countries.

	Size, 2011, US$, billion	Growth Rate
Global	694	3%
Brazil	23.7	7.10%
Russia	17.4	7%
India	13	11%
China	81	7.20%
Germany	22	1.30%
United States	133	1.30%

Source: Multiple market projection reports and authors' insights.

Figure 7.3. Global packaging market, 2011 data.

emerging markets, while its volume in Europe actually declined by more that 1%. The impact on packaging is that it has to meet the price restraints, yet reflect the quality position that Unilever seeks to reach.

Key action points for global markets

Growth is for the taking in developing economies

However, the rate may not be as rapid as forecast a few years ago.

Packaging Action: Watch the BRIC countries—Brazil, Russia, India and China. They have economies that will grow at least four times as fast as developed Western economies.

Technology remains a key factor in both developing and developed economies

Packaging Action: For growing markets, it can give them the edge they need to sell to export markets. For developed markets, technology can help offset the lower labor costs that give imported goods a cost advantage.

The consumer defines an economy's needs

Packaging Action: Packagers need to get good insights into each developing region's needs to succeed. Make products and packaging affordable; use the concept of "frugal innovation" to get there. Look for the social agendas; both government and non-government organizations have them and believe business should support them.

Build trusted supply base relationships

Packaging Action: In emerging markets, supplier qualification efforts may have to go deeper than in developed markets. It goes beyond reliability, quality and price. Government and interest groups may focus on social responsibility and building local economies. Supply strategies have to accommodate priorities like these.

Distribution tactics are essential for success

Packaging Action: Know the channels in developing markets—some areas retain traditional retail models that aren't well known in developed economies. For packaging, the last mile may be the most demanding for product protection. This is particularly true for foods where losses in distribution can be critical in developing markets.

8 || *It's a Risky World Out There*

> ***A scenario . . . On the Internet.*** Two almost contradictory stories cascade through the web. In one set of stories, consumer interest groups tout a report that a major food processor will phase out BPA from its cans. The second set of headlines say, "FDA won't ban the chemical BPA in food packaging." Expect these kinds of developments into the future.

Those two news threads on the Internet give us clear evidence that—today—packagers face risks from a lot of places beyond hard scientific evidence, regulations or legal action. Risks also come from groups that use the media and e-everything connections to influence a range of issues, including packaging. Add to that the risks from global counterfeiting and from global regulations. What it means is that risk management has changed for packagers. More than ever, it needs to be part of a larger strategic picture.

Let's look a little deeper into the Internet headlines on BPA. A major food processor became a target of the Breast Cancer Fund because of bisphenol A (BPA) in can linings. The Fund said that tests found what it called high BPA levels in a variety of canned foods, including products the Fund said would likely to be in children's diets. The group worked to target the food processor with its "Cans Not Cancer" campaign; the result was about 70,000 letters to the company asking the company to remove BPA from its cans. While this was going on, the U.S. Food & Drug Administration (FDA) was researching BPA in foods; it was responding to action by another non-government organization, the National Resources Defense Fund. The FDA research concluded there wasn't enough evidence now to ban BPA in food packaging. But, FDA also reserved the right to change its mind later.

This scenario reflects what every packager faces today—the potential for reputation and regulatory risks. And in many ways, every packaging manager deals with risk management. To get a wider look at best practices, we talked with Bob Pojasek. He's a Ph.D. who leads the sustainability consulting practice at The Shaw Group, Inc. and who has experience with corporate risk management. From his perspective, risks are often arranged in three categories:

- *Reputation risks.* Brands and companies face both consumer and media pressure that may range from mild dissatisfaction to outrage. They also face risk from counterfeit products that could diminish a brand's value. Ultimately, companies measure the impact in terms of financial costs and brand equity.
- *Regulatory risks.* They include laws and regulations. For packaging, regulations on chemicals and toxics are an important part within this category.
- *Operations risks.* Events that cause lines to shut down.

Some packaging issues cross into all three categories, and it is more convenient to put them into a functional segment. A critical segment encompasses fraud and its "niches"—counterfeiting, diversion, tampering and supply chain threats. They are reputation risks, yet they also have strong operations components and also include legal ramifications.

Risk management is a corporate function, and packaging is part of a total effort. In the broadest sense, the job is to array risks by priority. Pojasek suggests a "risk register" as a tool in assessing the risks. In essence, it is a database that identifies a risk, analyzes it and says whether it is acceptable or unacceptable. (For details and a sample form, see Chapter 10. Resources and References.) From a corporate perspective, acceptable risks can be managed; unacceptable risks that are essential to a business are insured.

When managers assess risks, they begin to see a priority order. Pojasek likes the Risk Map in Figure 8.1 as a way to illustrate the process. It parallels a number of models you can find, and it shows you where your projects stand in a relative order. Events at the lower left have a very low likelihood of occurring and a very low impact. They are acceptable. Think of a cap completely

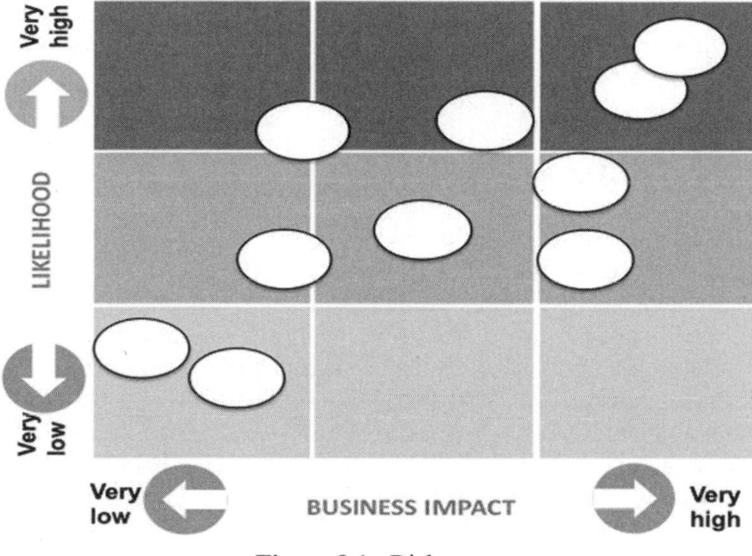

Figure 8.1. Risk map.

missing and a no-fill of the container itself. Current engineering practices make that highly unlikely, and the impact should be little more than a single irritated consumer. In the upper right hand corner are high-impact, high-probability risks. Think of a food packaging process with a high probability of introducing a pathogen into a container in a high-speed operation that contaminates a large number of containers.

All risks have a financial side, Pojasek continues. The mapping process highlights the unacceptable risks. Companies generally buy insurance to hedge against them. By lowering the degree of "unacceptability," a company can save considerable money on the insurance. The holistic approach to development we've outlined in this book is a way to address those risks in packaging.

Pojasek makes this observation on how businesses look at risk assessment: "One thing that I find fascinating is that there are two kinds of risk. We all know about the 'bad' risks. But there are also 'good' risks. These are opportunities to improve. Yes, they are risky. If a lot of money is invested in an opportunity and it does not work, the risk is then manifest. Many companies do not follow up on opportunities because they are risk averse." We discuss a company's ability to tolerate risk in the chapter on innovation; our observation is that a company with low risk tolerance may not lend itself to real innovation.

REPUTATION RISKS

We can look at reputation risks in terms of: "When the public casts its eye on packaging." Very often the risk is perception rather than reality, but as public relations pros say, "Perception is reality." Amplify any risk by the media's insatiable appetite for something to publish on a 24/7/365 schedule. And, social networking provides a less formal—but no less powerful—route to changing perceptions. If you stay around packaging long enough, eventually you hear about the Tylenol tampering event; it is the classic case study in how to manage a reputation crisis that revolves around packaging. It had deliberate tampering, fatalities, a massive recall of Tylenol products, and, ultimately, saving the brand's reputation.

Most reputation risks won't threaten to destroy a brand. But, as you look at your risk management map, consider what events may present a reputation risk. The first step in a strategic response is to find ways to mitigate the risk. Here's a place to use the Integrated Value Web as a strong tool to identify potential problems. In early 2012, the KFC unit of Yum! Brands saw this lesson play out. Greenpeace tested KFC packaging in global markets and said chicken buckets contained fiber from Indonesian rainforests. The source of the risk was a paperboard supplier that Greenpeace said was harvesting rainforest trees. That activity opened KFC to a reputation risk. In the chapter on green packaging, we note the need for a chain of certifications to verify a company's environmental position.

It's called risk because even after diligent work in protecting your brand and product, unforeseen consequences can still happen. One step that is increasingly becoming part of packaging's response is track-and-trace capabilities. The sophistication of coding techniques and parallel data process systems are becoming common operating procedures. These features may be built into packaging because of regulations, particularly with drug products. When it comes to an event, speed in accessing the tracking system becomes critical. Should the response require a recall, the tracking system helps assure that you recall all the product or packages need to address the event.

Finally, after you have taken all the steps to preclude, address and remedy a problem, it may become time to respond to the media. The response needs to protect your brand and your company. In assessing how packaging becomes part of a media response, we talked with Alan Isacson. He is CEO and founder of ABI, a marketing and public relations agency. He has a unique perspective that led us to him: he's a public relations pro who also knows the packaging community. We listened as he outlined critical steps you need to take when risk turns into packaging-based crisis:

- Tell the truth; be transparent. The e-everything consumer is skeptical of business and their antennae go up at the slightest hint of anything less than the truth. Isacson says you need to objectively explain what you did and why you did it. Be ready, too, to explain the pros and cons of the way you're handling the situation. Talk about what you did right and what you did wrong.
- Act fast. If you don't act fast, the stakeholders involved will began coming up with their own conclusions or altering the facts to fit their own perceptions. That may include special interest groups who have their own agendas to advance. Being first may "trump" communications from other sources who may not be accurate. Brief customers and suppliers without delay. Make sure sales people understand the situation and know how to explain the issue and answer questions. Everyone needs to be on the same page.
- Know who your audience is. Use language that's at a level they understand. Tell consumers about actions you're taking for their safety—that's what they understand. But be ready to give regulators details of how you are addressing their regulations. Other key audiences include investors and partners. Isacson emphasizes this group in particular: Employees. "They are your ambassadors. Get them on board first," he says.
- Be consistent. Most people can absorb about three to four messages, maximum. Identify those messages, then look at

your audiences and be sure each communication includes the key messages.

- Finally, have a communications plan. Often, this is the role of corporate communications. When it comes to packaging, you may be the only person who really understands the issue. The early plan should include tools to use with the employees and media to be sure the message is consistent. Know that your communications people are probably going to designate a spokesperson—that helps keep the message consistent. Know who that person is so you can direct calls to them.

Keep this in mind: You may not even be responsible in the event's genesis, but you have to react. A food processor had to recall cheese in flexible pouches because of a wild link to terrorism; someone suspected that a starch powder on the film—there to reduce clinging on rolls—was anthrax. The message was that the food processor valued its consumers' safety and it recalled the packaged product.

HOLISTIC PACKAGING ANSWERS FRAUD

Counterfeiting is the most visible form of fraud in packaged goods, and tampering, diversion and adulteration are other threats. Fraud even includes the converter who prints added units of a trademarked design and sells the over-run to a counterfeiter. Fraud is pervasive. The global economy makes it so profitable that some experts say there is virtually no product line immune from it. Even the cheapest items can be counterfeited and still deliver a profit to those who make and package fake products. As a ballpark number, counterfeiting is an "industry" with global sales ranging from US$ 200 billion to US$ 600 billion. Look at just one industry—fake auto parts cost manufacturers about $US 12 billion globally. Think, too, of brand protection. Contaminated milk in China eroded the market for local dairy brands as consumers saw national and international brands as answers to quality concerns.

Whatever the fraud threat, packaging can be part of the solution, but it takes a strategic approach to have an impact.

To get a better scope of the strategic approach to product

fraud prevention, we turned to John Spink, Ph.D. He is the Associate Director of the Anti-Counterfeit and Product Protection Program at Michigan State University. Spink puts the solution in terms of "holistic thinking." Doing that takes us into some territory that is not usually within the realm of packaging—particularly the view of the crime's nature and motives of the people doing the counterfeiting. It's a human action, so we find ourselves in behavioral sciences and criminology. And, being there helps formulate a plan.

On the scale's lowest end are the recreational and occasional counterfeiters. Going up, we meet occupational, professional and ideological counterfeiters. The last group includes the global terrorists who make an ideological statement; they may try to do economic harm with fake products. Whatever the level, the avenues they use are shown in Figure 8.2. The impact of these threats can range from public health issues to economic losses.

There are a number of international guidelines on product protection practices, including ISO guidelines.

Here's where Spink sees as packaging's main role in the process. "It can help detect any of the threats. In counterfeiting, for example, packaging design can help detect counterfeits, but that is not a single 'magic bullet.' Packagers have to look at the value chain including manufacturing, inventory, distribution, retail channels, the consumer and even disposal." His comments parallel our belief that knowing the integrated value web is a key step in any aspcct of packaging, including product fraud. Figure 8.2 shows some of the routes to product fraud.

Adulteration	Using a fraudulent component within a finished product
Tampering	Altering a legitimate product
Over-runs	Excess production of a legitimate product or package diverted into fraudulent channels
Theft	Stolen products
Diversion	Distribution of legitimate products into unauthorized channels; sometimes called the "grey" market. It may violate laws or a supply agreement, but it is not explicitly defined as fraudulent.
Counterfeiting	A fraudulent product and package designed as an exact replication of a legitimate product.

Figure 8.2. Fraudulent practices.

The first step is to assess the situation. Ask where your package and product could be compromised:

Your suppliers and their suppliers. Where can fraudulent components be introduced into the supply chain? Do you know where your suppliers get their supplies? Do you have trusted suppliers and long-term contracts? Do you trust but verify?

Your distribution channels. Key actions within channels include track-and-trace capabilities in packaging so you can know where your products have been. Often, that means a "feet on the ground" in international markets and the capability to verify the authenticity of packaging. Here's where packaging comes in—a number of authentication methods help people in the field confirm real packages and spot fakes. Spink cites an axiom of brand protection experts: "If you haven't looked for counterfeits, then you don't know that you're not being counterfeited."

Answering a threat to a brand's integrity means looking at all possible points of product fraud. Figure 8.3 represents a simplified, generic concept; however, each member of the value chain needs to develop its unique fraud map to identify weaknesses. For example, an error at a raw materials supplier can impact the brand owner, a retailer in an authorized channel and the consumer.

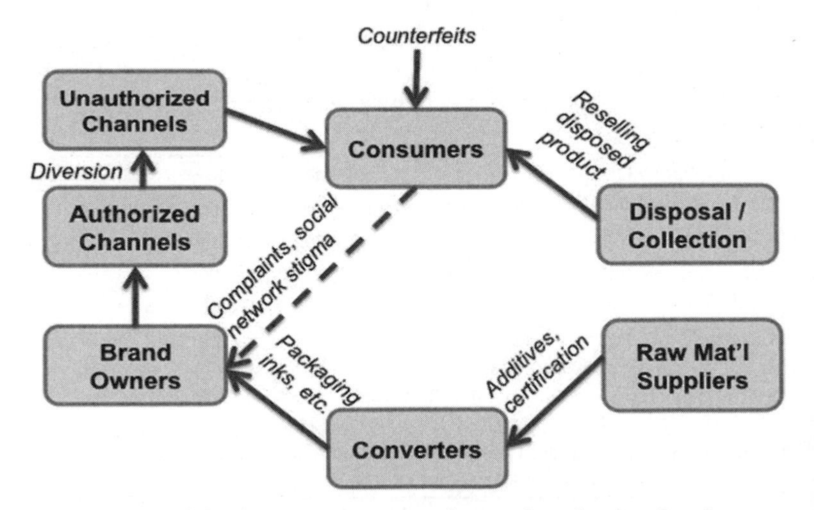

Figure 8.3. Some routes to product and packaging fraud.

REGULATORY RISKS

In addition to threats to their reputations and brand equity, packagers also have to deal with packaging regulations, and the major ones have global impact. Those include regulation of chemicals and toxics-in-packaging; they most frequently affect additives and components in both packaging materials and components such as adhesives. The emphasis on "green" and the demand for healthy products makes the public more aware of potential problems. Often, addressing these risks means working closely with other members of the integrated value web and keeping up-to-date on developments.

The major regulations with potential to affect packaging are the U.S. Toxic Substances Control Act (TSCA) and the European Registration, Evaluation, Authorisation and Restriction of Chemicals (REACH) authority.

Overhaul of TSCA has been a U.S. Congressional goal for several years. However, legislation has moved slowly. Even if there is no U.S. federal legislative move on TSCA, regulations may pose risks. The Obama administration, in mid-2012, sought to move chemical management up on U.S. Environmental Protection Agency's agenda.

REACH is a European Union initiative that went into force in 2007. Its scope means that implementation continues to evolve. That involves both registration of substances with the European Chemicals Agency, and further implementation by member states. As of 2012, the highest likelihood includes some phase-outs or restrictions on candidates of high concern and a look at recycled plastics. Added regulation of inks, adhesives and paper are other areas that bear watching. The inclusion of recycled plastics is another focus on regulatory efforts.

Toxics in packaging

A state-level initiative in the United States, this effort aims to reduce heavy metals in packaging. Almost 20 states have adopted model legislation, and the Toxics in Packaging Clearing House supports efforts with promotional efforts.

Nanotechnology

Nanotechnology has drawn interest of government and non-government organizations concerned with health risks. On the legislative side, the Nanotechnology Safety Act of 2010 would have given the U.S. FDA regulatory oversight on this technology; the legislation died in the U.S. Congress. Among NGOs with a nanotechnology agenda are Greenpeace and Friends of the Earth. Impact could include additional scrutiny of safety over the next several years. Another possibility is prohibition of nanomaterials in packaging for organic products.

It is beyond the scope of a book like this to develop a comprehensive list of the status of laws and regulations on packaging. However, companies in the field can develop their own monitoring programs. Consider these steps in developing such a program.

- Know what chemicals pose risks to your packaging, and be alert to the possibility that they may the target of regulatory action. If they are likely to be a target of regulatory change, find alternatives.
- Look at your exposure. BPA is a primary example. Consumer pressure has driven some packagers to find alternatives.
- Keep your horizon wide—the U.S. federal government isn't the only jurisdiction with high potential impact. If you are in the United States, state level efforts—particularly California—could change the game. Europe's REACH regulations could also have an impact.
- Look at chemical management initiatives of players within the packaging value web. They are government and non-government organizations that are in the influencer positions within the web.

Key insights to addressing risk

Be prepared by looking at future scenarios

If you examine what could happen, you are better prepared if it happens.

Packaging Action: Tools like the PTIS foresight scenario process let you do your own foresight work to spot potential risks. It helps you probe potential pitfalls in detail. With those insights, you are more prepared to act quickly.

Move toward formal risk assessment thinking

Know that formal, corporate-wide risk management processes can help you define and rank the risks you face.

Packaging Impact: Use tools such as a "risk register" to name, define and assess the risks you face.

Reputation risks include hard-to-define "perception" issues

Groups such as NGOs may target your company, or some process your company uses as part of their mission.

Packaging Impact: Make sure your processes are on solid ground, and then be ready to be part of your corporate response to any issue.

Fraud is pervasive in a global market

It ranges from counterfeiting to diversion, tampering and adulteration.

Packaging Impact: Know where your weak points are, then use a holistic approach to develop strategic answers.

Regulatory risks may become more common

Consumer drives toward health and wellness extend to the packaging for products. Government and NGOs will identify and attack problems.

Packaging Impact: Know where those risks might be by monitoring regulations and NGO activity for materials that are "mission critical" to you. Have alternatives ready.

9 || *Strategic Thinking, Innovation: Packaging's Pivot Points*

> ***A scenario . . . Sao Paulo.*** The cookie package's bright graphics delight the pre-teen and his mother as they ask the store clerk to put it in their shopping bag. What they don't know is that innovation by a Brazilian film converter five years earlier brought them the home-compostable package. It resulted from an innovation program that assessed and weighed risks against benefits. First, the converter looked at "green" regulations and guidelines in Brazil and saw the best position was to anticipate changes. Then it worked with local cookie makers to get insights on consumer perception in the category. The converter tapped into a knowledge base it had to find plant-based materials that made compostable films feasible. The converter's business plan also factored in the export market where the film's increased value offsets the longer lead-time in shipping film to the north.

Take all the strategic components of packaging development we've examined in this book and tie them together. What you have is an action plan that turns an idea into a marketplace reality. Done right, the process delivers value to the consumer, including the business-to-business customer. It complements and supports the benefits of the product the package transports, contains and promotes. It is the practice of the value formula we detailed in Chapter 1 and amplified in the appendix:

$$\text{Value}_{\text{buyer}} = \frac{\text{Product}_{\text{Benefit}} + \text{Package}_{\text{Benefit}} + \text{Experience}}{\text{Price}}$$

In the context of integrating all of the parts into a whole, packaging innovation becomes a core business function. It delivers a sum that is greater than the value of all the parts. The process starts with a company's overall business strategy as input, and it yields a package/product combination that supports that strategy. What we've done in this book is touch on all the packaging components that feed into the value formula; and we've showed that—in many cases—the product and experience factors are inseparable from the packaging components.

For leaders like P&G and Kraft, packaging innovation has grown into best business practice. Yet, the benefits of practices like those elude many businesses. A recent PTIS report points out that just 25% of business leaders say their companies are successful at innovation. More data from the same report: 70% of executives say that innovation is among their top three corporate priorities, yet two-thirds of them are not confident in their ability to execute innovation. To do that, each company has to adapt a working innovation process to its unique goals, market, supplier relationships, customer relationships, and culture. What we are going to show you are the key building blocks for innovation and let you determine how they can help you build an organization for innovation that fits your company's needs. The strategy is written primarily from the perspective of the consumer packaged goods company; the template, though, easily fits retailers with a strong private brand strategy. It also fits those marketing packaged products in the business-to-business arena. And, although our examples are of CPG projects, the process also applies to suppliers along the packaging value chain.

THE ROADMAP TO MANAGE FOR RESULTS

The steps that put packaging innovation at the pivot point of any business are actions that integrate packaging into a business strategy. From our research and from projects we've done, we see nine building blocks that help make packaging a solution. The process fits a broad spectrum of businesses, yet details can differ significantly for individual companies. Here are the Big "P" Packaging steps we have seen replicated in businesses with successful innovation programs:

1. Know your organization's strategic business drivers

We've all worked with some of these drivers—increased productivity, better margins, growth in market share, expansion of the distribution network, and cost savings that don't have a negative impact on the consumers' brand experience. We've talked about sustainability earlier, and for leading-edge companies, it joins some of the more traditional drivers. Other drivers include brand and product differentiation. Some businesses state innovation specifically as a driver, having objectives such as developing a target percentage of sales derived from innovation. It may take some work in pulling the drivers out of mission and vision statements along with three-year strategic business plans; sometimes, you may find insights on drivers in business unit operating plans. Knowing the specifics of the drivers helps shape a packaging innovation strategy.

How will you know you're connecting with the drivers? Our experience says to look at how your organization integrates customer and consumer insights into the development process. Is there a stage in the process where the team compares consumer insights and business drivers side-by-side? Does the development team analyze the fit, to see if insights and drivers complement each other? If so, and if there are realistic discussions at that point, chances are that innovation leaders understand the strategic business base for action. Best practice companies have multi-year integrated business strategies that link brand, product, packaging and capital plans. Most often, we see those strategies in a three-to-five year timeframe. Some push the window to 10 or more years. These strategies are supported by cross functional innovation teams; they are charged with identifying and connecting new materials, processes and knowledge to deliver known and anticipated need states.

Another key element innovation leaders have to understand is their organizations' risk-tolerance DNA. At the leading edge are the acknowledged risk takers; these companies will take on the venture to find disruptive innovations. Not far behind are the fast followers. For both types of organizations, there is a strong emphasis on connecting with the customer or consumer. These are the innovation leaders. Further along the spectrum are the "incremental line extension" operations and the "me toos." It is OK if your company is one of those, but chances of implementing a wide innovation program are slim.

2. Identify barriers to innovation

Here's what we hear people most frequently say are barriers to packaging innovation: Cultural resistance, red tape, organizational structure and the "not invented here" syndrome. "It costs too much," "the timing isn't right," and "we don't have the resources" are other barriers people talk about. A lack of focus, or lack of an innovation strategy, is another barrier. A final obstacle is the lack of a way to quantify value; you have to prove value to have meaningful innovation. If you see those barriers popping up frequently, expect that packaging innovation will be channeled into paths that more often produce incremental changes rather than true innovation.

Some of the subtle versions of the barriers include a business structure that follows traditional, linear business planning models. The traditional models rely on input from obvious, traditional sources; they often follow articulated consumer needs rather than the unstated needs. Chapter 2 contains a list of traditional consumer needs; these are the articulated needs, and they are really a point of departure. If an organization stops there, it is an impediment to innovation. Does your organization use tactics such as ethnographic research that uncover unarticulated needs? If it does, it has stepped past a significant barrier. In worst-case scenarios, a company ignores consumer insight research and becomes overly reliant on new technology itself, seeing it as a "silver bullet." Facing such barriers, any well-intentioned formal innovation program is a struggle at best.

3. Detail opportunities and problems

Look at opportunities and problems in terms of detailed business needs. We've listed some of the more common drivers, but how do they translate into opportunities? How do you translate them into solutions that address customer and consumer needs—convenience, health, wellbeing and more. We think Method Home Products offers a model; its business drivers include a strategy to grow through added product offerings and gaining market share in its current categories. It also wants to answer consumer needs for sustainability in products. By marrying the drivers with consumer needs, the company's 8X concentrated detergents succeeded in the marketplace. Consumer research also uncovered unarticulated needs about messy detergent packages. Method turned the household-cleaning category upside down.

4. Have a portfolio of projects

The portfolio concept says you have projects in the short, middle and long-range timeframes and with big, medium and small impacts. When you do that, you can give management a flow of successes. Management wants to see a return on its investment; if everything is a big, long-term project with a five-year payout, the risk is that management's patience wears thin. A portfolio delivers a probability that your flow of successes can show the value in innovation. The grid below shows a small tactical change in the lower left hand corner; it is a graphic change to a package and it requires few resources. In the middle right hand column is

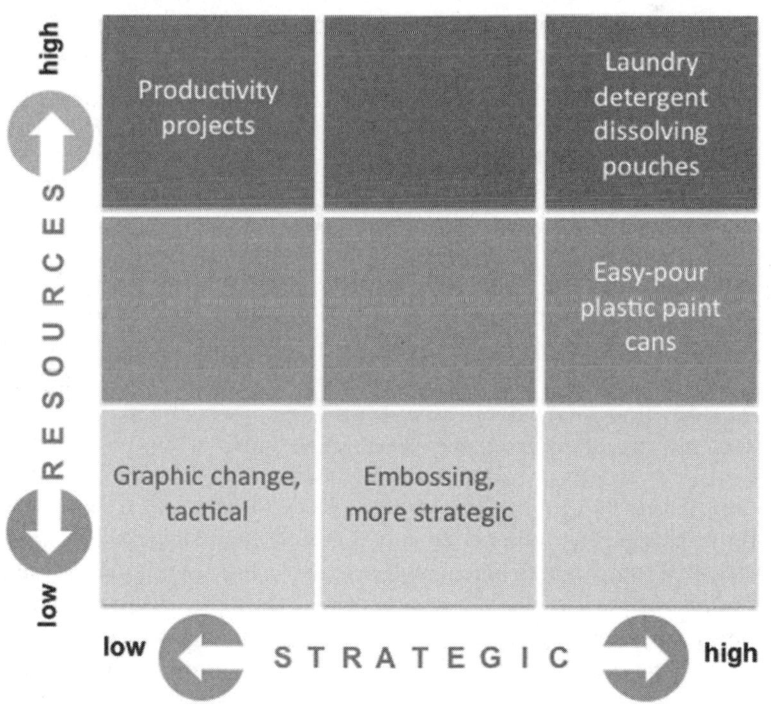

Figure 9.1. Grid for portfolio management.

an innovation that includes the consumer research and a significant variation on the container style; yet, it retains the same basic filling equipment with some modifications. In the upper right hand corner is a new packaging system that is strategic; it answers an unstated consumer need, opens up a new niche within a category, and requires a significant commitment of resources, particularly investment in new packaging equipment. However, it offers a potentially large payback. The portfolio concept says have projects that cover the range of options.

5. Use the value web as a management tool

The integrated value web detailed in Chapter 4 is the most strategic tool for innovation. Be ready to use the external components and have cross-functional teams in place to implement projects involving the value web. With the pace of innovation up to expressway speeds, you have to know the map now, rather than figuring it out as you go. How can you gauge if the process is working for you? Ask this question: "How well do your internal, cross-functional teams perform?" If they deliver the ideas and move toward goals, then chances are they are using the integrated value web effectively.

6. Sharpen your focus with future scenarios

Innovation is about the future, and scenario development may be one of the best tools to help you better estimate the future. Scenario building is more than forecasting. It looks beyond what traditional, straight-line management tools foresee. In scenarios, you develop alternate futures rather than extrapolate from the present. Here's an example: create a scenario around your most heavily used packaging material—what happens if the price doubles or triples?" Look at the additives in your packages—build a scenario around what happens if one of them has such a bad public image that using it reduces sales. Done right, you build a system that evaluates alternate futures, challenges and opportunities for packaging. In addition, scenarios shape a continuing "strategic conversation" on how the future impacts your business plans. They are foresight driven. Creating scenarios is a multi-step process:

- Step one is to identify a focal issue. As an example, let's envision a company with a strong emphasis on sustainability with products whose value proposition is built on a friendly-to-the-environment concept. A focal point could be on materials that will be in the company's packages in ten years.
- The next step is to list factors and forces in the business environment that impact the focal point. Bio-resins development is one factor to consider, including development of bio-based raw materials. Consumer attitude is another factor. Technical development in other eco-friendly materials such as paper or molded pulp can be other factors.
- Then, identify critical uncertainties. For example, a "crops for people" trend could change raw material availability and pricing. The scenario-building process ranks the uncertainties and creates a grid. You also have to consider safety risks in nanotechnology; already, non-government organizations are looking at ways to assess those risks.
- Those steps create a grid, and the next action is to "flesh out" the scenarios. This process leads to implications and options. Those implications should answer questions such as "What would it mean to the company?" "What would it mean to relevant stakeholders?"
- A critical step is presenting the scenarios to management. Each management style will be different and suggests ways to get the information across. Graphic presentations often have a high value because management is visually attuned. Consider, too, that the presentation may have to stand on its own for subsequent "showings." You can follow the grid to present a logical sequence.
- Future scenarios develop options, not the "truth." They make your company more prepared to deal with possible changes, but they do not define the course of action. If an event triggers the need for action, it may involve a company-wide process where a number of functions bring their "truth" to the table. What packaging has provided is the enabling jumping-off point.

7. Have the right people in place

The right person is a generalist who also has to be a strong champion of packaging's value. The title may be Chief Packaging Officer, Director of Innovation or Manager of Packaging Development. The person's skills have to embrace technology and still work effectively within an organization. One factor we're seeing as we did research for this book is the growing awareness that packaging is a social tool. Good packaging managers have to be adept at understanding consumer insights and needs of society in which the packaged product is sold. Here are some skills criteria we suggest:

- Does the person have good team-building skills, especially when unconventional people are involved? Good innovation builds on good ideas, and the teams around packaging processes usually include people with different ways of seeing things. A creative designer is going to see a challenge differently than a technician in the lab. The leader's skill is to channel their efforts into products that deliver value.
- Networking skills. The right leader will sometimes be a "technology scout." At other times, a "librarian" for information on consumer preferences and habits around a particular product. The packaging leader has to rely on a network to sort through the right answers.
- Adaptability to lead in uncharted area. Innovation, by definition, takes teams into activities that aren't part of a company's normal operations.
- Drive to finish. Tenacity is a special value when major assumptions underlying a project may shift significantly half way through. Is the packaging leader comfortable with ambiguity? Can she avoid a pre-set conclusion on where an innovation initiative should arrive?
- Communication skills. Can the person take all the diverse components that shape a project and turn them into a cohesive story that management will understand quickly? Does the person understand that it is not about "water vapor transmission rate," but rather it is about "extended distribution channels"? Can the person walk around with the "elevator speech" in her head so that a chance, 10-second conversation with a senior manager is productive in getting an idea across? The elevator speech needs to impart knowledge.

Think in terms of what is the project's status. What have we learned from research or technology scouting?

8. Don't let the "crisis mode" swamp innovation

Everyone who is involved in packaging is going to spend time "putting out fires." Yet, we have to make sure the innovation process unfolds and coexists with those priorities. The innovation leader needs to articulate the innovation strategy clearly and strongly. Management has to know just how important it is to support innovation along with day-to-day fire fighting. Management needs to understand that if they cut short development time, one result is the need to take more time later to fix the problem.

No big surprise, but in the real world the amount of time spent fighting fires—and the impact on innovation—clearly differs from the ideal. Here's some proprietary, "best in class" research from Packaging Technology Integrated Solutions. In analyzing top consumer CPGs and packaging suppliers, PTIS found that more time is spent fighting fires, and less on innovation, than what the industry considers the best ratio. Management's role is to try to shift effort from fire fighting to innovation.

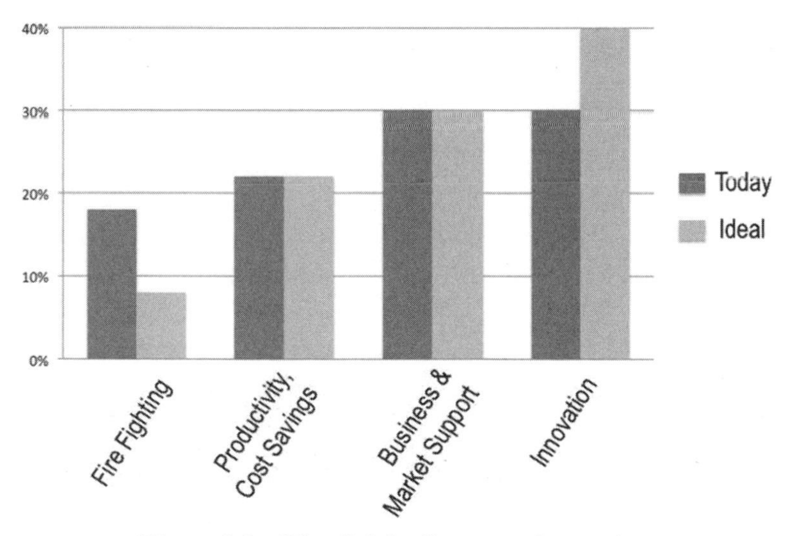

Figure 9.2. "Fire fighting" swamps innovation.

9. Develop the right metrics and know when to kill a project

Business is about results, and they need to be quantifiable. Build those metrics into the operation. For example, a metric might be a measure of how much revenue new products account for in the company's overall revenue picture. They might include a look at product successes, or failures; a company may establish a metric that it reach a 35% new product success rate, determined by how many new products are still on the market three years after launch. The metrics keep the innovation process focused and, if goals are met, demonstrate the return on the innovation investment. With the right metrics, you also have a better rationale when it is time to kill a project. The ability to do that is a healthy indicator of a successful innovation process. If you've set metrics to measure the lack of progress, it makes it a neutral business decision rather than a negative or black mark on the program leader or project team. It means that the innovation direction doesn't work against the established criteria and metrics.

Checklist. The roadmap to innovation results

1. Know your organization's strategic business drivers
2. Identify barriers to innovation
3. Detail opportunities and problems
4. Have a portfolio of projects
5. Use the value web as a management tool
6. Sharpen your focus with future scenarios
7. Have the right people in place
8. Don't let the "crisis mode" swamp innovation
9. Develop the right metrics

OPEN INNOVATION

A number of innovation models dot the business landscape, and the criteria outlined above embody the thinking behind them. One of these models is Open Innovation, and it is the predominant approach for packaging innovation for consumer packaged goods companies (CPGs), or, if you prefer, fast moving consumer goods (FMCGs) companies. These companies serve wide markets and rely on a range of technologies to do that.

How innovation plays out in the real world

Neil Darin is an innovation maven. He's a Senior Program Manager for Innovation and Sustainability at HAVI Global Solutions. What Neil brings to the table is a perspective gained in working with innovation programs for a Fortune 100 consumer packaged goods company. We asked Neil this question, "From your experience, what makes innovation work?" Like so much we found in putting this book together, it goes back to the basic issue of how innovation fits into a company's business strategy.

"Innovation is really about targeting where business opportunities are," he explains. "You have to start with your company's business strategy—what is the SWOT analysis? What are the quality challenges? What are the competitive strengths and weaknesses?" Darin goes on to list more criteria including insights on consumer marketing and regulatory trends. Finally, he says, ask what problems you can solve related to your situation by leveraging technology and design trends.

If you ask these and other questions, you get to a point of identifying what Darin calls "Opportunity Platforms." They emerge where business drivers, trends and regulations, and technology overlap. An opportunity platform is a solution that is specific to a market, an industry, and a company. Here's a way of looking at it graphically, and the opportunities emerge where all three circles intersect.

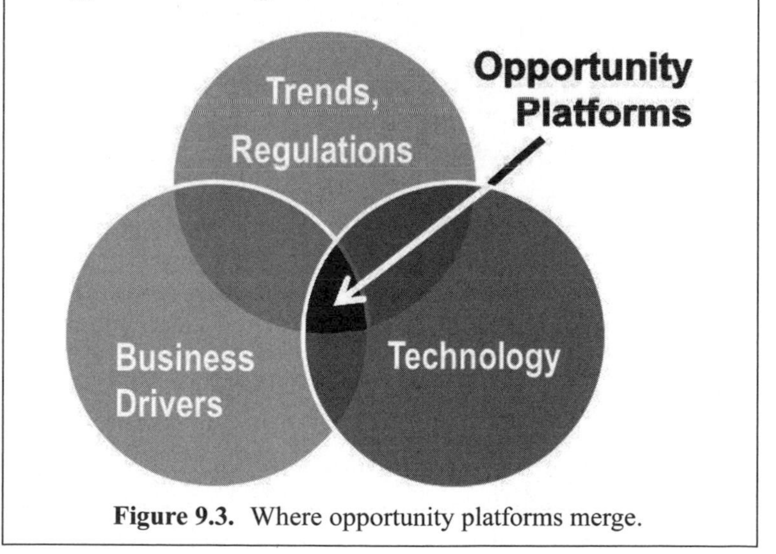

Figure 9.3. Where opportunity platforms merge.

Consider multinationals such as Kraft, P&G and Nestle where product lines and packaging span a range of platforms. Finding the right one is often the biggest challenge in package and product development. Open innovation creates the two-way street to search for answers; it can be quicker and less costly to travel than the limited path for internal development. Innovation also gains from the unique brand power that large, multi-nationals have. Rather than having to rely solely only on the innovation's benefits, development can gain from the marketing value of strong brands. For small companies that own technology, that can be a real plus in commercializing an innovation.

The core strategy behind open innovation is simple: *Look outside your company for ideas, and not necessarily from traditional external sources.* Suppliers are a traditional external source of ideas, but open innovation suggests going further. The integrated value web is the "turf" on which to search for ideas. Among emerging sources, look to university think tanks; small start-up technology incubators are another. Searching the value web, and looking at these kinds of organizations in particular, opens wider pathways. In particular, it gives companies routes to global technologies that may not have been "on the radar" before. Figure 9.4 outlines differences between traditional closed innovation and the open innovation concept.

The basic concept is clear. Yet in many cases, the promise of open innovation has not been filled. Often the difference between gaining the benefits and missing the mark are a few obstacles that stand in the way of the collective approach to gain new ideas, packaging and processes.

Open Innovation	Closed Innovation
Good talent resides beyond our "four walls".	We have the best people internally.
Technologies and ideas reside among non-competitors. Even when we deal with competitors, we can have win-win situations.	To win, we need to discover it first and hit the market first.
Can we buy innovation?	We keep control of innovation so competitors can't "knock off" our ideas.

Figure 9.4. Comparison of open vs. closed Innovation.

Failure to build trust. The first need is to build trust among collaborators. This is a mandatory ingredient for open innovation to work. The corollary is to have good business agreements and contracts to support trust. A secondary "must" is that an organization has to be willing to accept outside ideas. Often, it takes top management to foster and build that kind of thinking—be sure it is in place with your senior management and with your prospective partner's top management. Without those both in place, all the other efforts around open innovation will be less efficient. If either is low on an organization's list of priorities, then open innovation efforts may be ineffective.

There may be just a few trusting business relationships with partners. Some will say there is a "gut feel" when those relationships align. Here are some questions to ask to reinforce that "gut feel." How well do the cultures of each partner align—how do the goals, objectives and corporate cultures match? One example: To what degree do both organizations work with cross-functional teams versus silos? If they are at opposite ends of the team/silo spectrum, chances of success are minimal. Another question to ask: Is a commitment to satisfying the consumer or customer a crucial goal? Our experience is that disasters happen when one party glosses over a customer or consumer obstacle. There is no place in the relationship for thinking like this: "Marketing can take care of that later," or "Logistics will figure out an answer." In a holistic approach, you have to include all the potential outcomes and have at least a conceptual answer in the early stages. Rarely do those "loose ends" get any better as the project goes forward.

Not seeing the win-win situations. Can the process lead to a win-win position? That's the next step once you know you have a general business fit. Both parties need to share ideas openly and share the results together. Alliances have to offer significant value to both sides of a relationship. One party's strengths need to fill another's gaps. Here's an example: A university resource has a commitment to bring a sustainable material to market quickly, and a material producer also wants to bring sustainable innovation to the market soon. One has technology; the other has marketing capacity. Other complimentary capabilities might be relationships with equipment manufacturers, package suppliers and brand owners.

Failure to identify opportunity platforms. Open innovation's nature can let you wander further afield than closed innovation. You have to take conscious steps to keep open innovation "on target." One way is to identify the opportunity platforms that are most important to your company and focus your efforts on them. Later in this chapter, we examine why P&G's open innovation process works as well as it does. One factor is its decision to focus on specific areas. To find what platforms work for your company, ask, "What are the core competencies my company needs to have to stay competitive?" Do it in terms of your business objectives. It will help you see that some of the new ideas offer better synergy than others. What works best for you depends on your industry, product category and corporate objectives. For example, if you are a beverage packager, the core competencies might be marketing and delivery of beverages to consumers through specific distribution channels. That can mean glass and plastic bottles and pouches as platforms. For a different company, the packaging requirement might be to protect fragile objects; the platforms could be paperboard, rigid plastics or other formats.

Identify platforms by first defining your end needs. If you are too "tight" in your definition, then you may build platforms that are too narrow. For a beverage company, the end need is not how to produce lightweight bottles, it is how find the right value proposition that works within distribution channels and still delivers shelf impact and consumer convenience. Don't let your current production processes limit your answers; doing that may preclude consideration of a disruptive technology. One result of relying on internal legacy processes and equipment may be inefficiency. Your horizon could cover incremental changes and possibly miss discontinuous solutions with a big payback.

Not "nailing down" who owns intellectual property. Critical for open innovation is agreement on who owns what intellectual property. Spell out, too, responsibilities of each partner, accountabilities and an exit strategy. Open innovation says you have up front discussions and skeleton legal agreements on these issues—they are at the core of the relationship. Detailed legal agreements later will document the process, but you need to understand dimensions and positions first. Small points that can be big issues within agreements include termination clauses. And, be sure that intellectual property ownership survives the termination of the contract. In dealing with universities and research incubators, you need to know what they own, what they can sell, and what they need to publish. Subscribe to the principle that the relationship strengthens, and not weakens, each party's core business.

Not having clear definitions. Finally, you need to have clear project and program definitions. That includes timelines and defined deliverables. What are your project management tools? Some would consider a form of a Stage Gate process mandatory for open innovation projects. It is a common process, and we've put key elements of this style of management tool in the appendix.

The overall tone of the open innovation guidelines above says, "search the integrated value web for answers," and then lay out a way to do it that benefits all the organizations involved. It is a complicated process and moves the packaging professional into the role of being a strategic manager. Is it worth the effort? We're going to look next at how it works at P&G and the benefits they've seen from the effort.

WHAT DOES IT TAKE TO BE WORLD CLASS IN PACKAGING?

We've touched on key practices that help companies reach innovation goals. We're going to drill a little deeper and look at the corporate competencies that make it possible to reach those goals. We looked at some proprietary research by PTIS that lists 26 competencies; the list emerges from a decade of benchmarking by the PTIS staff. The competencies break across three platforms; here are the platforms, and some of the specific competencies.

- Organization and culture platform: The most important competency is a culture that fosters creativity. A sense of urgency and excitement is another key competency along with a commitment to sustainability.
- People and skill sets platform. In best-in-class companies, R&D people have business and communications skills, and manufacturing and QA get involved early in the develop-

ment process. Also important is having a staff that understands the consumer and is willing to share knowledge.
- Process effectiveness platform. A key is a process that is transparent, allowing communication among all functions involved. In addition, the process needs to confront obstacles that arise.

Within that framework, let's look at one set of charts that show how this kind of thinking is a strong management tool. Figure 9.5 lists five of the 26 competences for a hypothetical company. In this example, the dark bar to each cell's right is the nominal value derived from PTIS analysis of best in class companies. The lighter grey bar to the left is the actual level assigned after analysis of the company. The gap suggests where the company has specific strengths and where it might focus efforts to improve.

Figure 9.5 shows a hypothetical company and looks at just a few competencies, yet PTIS has applied this analysis to a number of major CPGs to identify the best and to use that information to benchmark other companies. The benchmarking process involved more than 100 thought leaders from academia, industry leaders and the PTIS Expert Network. Companies such as Apple, Method, Coke, Kraft, General Mills and Frito-Lay rose to the top of the list. They are not the only companies near the top, but they give you an idea of the caliber of management and packaging innovation that produces marketplace winners today. Not surprisingly, P&G emerges as best in class. Its organization, culture, staff skills and process effectiveness set standards. The fact that its new product success rate hovers near the 50% mark adds "icing to the cake". Here's an in-depth look at some of the key factors that drive P&G.

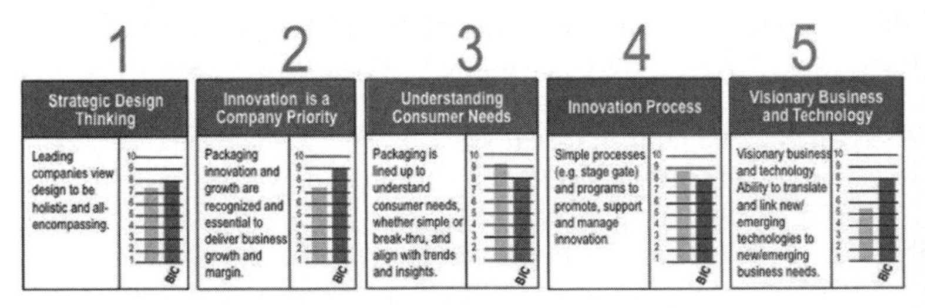

Figure 9.5. Analysis of competencies.

P&G'S OPEN INNOVATION MODEL STRESSES 'HOLISTIC'

P&G's innovation model certainly works in terms of the competencies, and the package development process highlighted in this book shows that it works well. We go a step further and ask the question, "Does the model offer a bellwether for the next decade?" We think it does, if for no other reason than the number of times P&G uses the term "holistic" to describe its process.

P&G has been open about how they organize, and its people have talked about the innovation management process often. They all echo this thought: Innovation can be managed as a holistic, disciplined, reliable process. We can add that P&G also seems to believe that sustainability is a core segment within the holistic process. To P&G, the most important stakeholder is the consumer. People within P&G echo that by saying, "Innovation is always about the consumer. The consumer is Boss." It is about reaching more consumers, in more parts of the world and more completely. The company focuses on delighting consumers from the moment they hear about a P&G brand, to the moment they choose it at the store shelf, to the moment when the consumer uses and experiences the brand. No wonder P&G is closely associated with the Moments of Truth; the strategy they outline simply encompasses the entire range from Zero to Third Moment of Truth.

The P&G organization makes sure consumer research translates that into packaging design by leveraging the P&G's First Moment of Truth staff. The marketplace payback: P&G notes that its loyal consumers buy brands more regularly and more often, and that they are less price sensitive. We think one example is the Olay Regenerist brand in China, which is packaged in the same upscale format it is packaged for the U.S. and Western cultures. It appeals to upscale Chinese consumers who are driving the emergence of a domestic market for Chinese production.

Since 2000, when P&G put the emphasis on open innovation, it says it wants the process to do two fundamental things: First, to improve the quality of life for people in every part of the world. Second, consider innovation as the primary driver of business, financial and economic growth. P&G itself uses the term "holistic innovation." It translates that into looking for innovation at every point the brand touches the consumer. Here are some of the key components P&G says are part of its holistic process:

Leadership. P&G expects it in its labs, in universities that collaborate with it, and at the senior levels within P&G itself. Leaders must see innovation as consumer-driven and must manage it as a social process. To get there, the company defines innovation broadly, yet makes it systemic—a process P&G can reliably replicate anywhere in the world.

Focused effort. P&G is a best practice in focusing efforts on its technical competencies. They have a deep technical expertise in about a dozen sciences and technologies, and use that depth to transfer ideas across business units. P&G reports better use of staff time by focusing its efforts.

Open Innovation. P&G stretches beyond its own deep competencies by looking for ideas outside the company. It is a classic study in open innovation. P&G's "Connect + Develop" effort is a visible part of its process. Call it "crowd sourcing" if you want, but it is the outreach that brings in ideas. The Internet portal is just one part of the outreach to gain outside technology and to share P&G work that can benefit other organizations. It is driven by the goal of having half of the company's innovations come from outside sources.

Design. Within P&G, it is called the Clay Street project. It uses a team whose goal is to find a way to imbed design into the company's brands and DNA, and it follows the lead of former CEO A. G. Lafley, who has said that design is as important as the materials that go into a product. Typical of P&G, the project also maintained an unwavering focus on the consumer; although the Clay Street project moved some traditional marketing functions to a cross-functional team, this emphasis remained: How does it "play" from the perspective of the consumer?

Stage Gate Process. If you look at what P&G has said about its model, it frequently points to the importance of the Stage Gate® approach. The company calls its version the SIMPL process (for Successful Initiative Management and Product Launch). A quick diagram of the P&G approach is in the Resources and References chapter along with a quick outline of the Stage Gate process.

Beyond the key components, P&G divides its innovation process into four major phases.

- It calls the first "Search and Discover." It parallels our

thoughts on leveraging the Integrated Value Web. It looks for ideas from consumers, retail customers, suppliers and other partners. P&G stresses that it looks across and beyond its own industry segments, and it looks in every part of the world.
- Next, P&G moves to what it calls the "Select and Resource" phase where it allocates human and financial resources to ideas that fit its goals. The phase also kills more than half the ideas originally presented, a step P&G says is a key to the process.
- "Design and Qualify" is stage three. P&G's process is consistent with our stress on teams. It uses multifunctional teams from production, research, marketing, manufacturing, engineering, finance, design and other functions. They bring the "holistic" view to each development.
- "Launch and Leverage" is the final step where P&G moves ideas into its product launch pipeline.

A couple of other aspects of the P&G process:

- The cross-functional team concept elevated technical input. Packaging technology moved in on a par with product technology. That structure produces concepts like Tide Pods, where the packaging and the product are one and the same. It also brings material technology into the design process.
- P&G people talk about "integrating innovation into its business." Doing that takes purpose, goals, strategies, strengths, structure, systems, leadership and culture. Yet, P&G makes sure that the process doesn't overwhelm the goals and that it is not reduced to a numbers game. It emphasizes that it is all about bringing value to consumers.

P&G has a best practice success rate in product launches, and its innovation strategy contributes to that. Estimates say the P&G succeeds with 50% of its launches, compared to an overall average of between 10 and 30% for consumer packaged goods.

TOTAL COST OF OWNERSHIP

If you take the holistic development approach, then you have also laid the foundation for an economic analysis called "total cost of ownership," or TCO. They compliment each other because the holistic path brings the entire process in focus so you can begin to make total cost estimates. This is where cross-functional teams really add to the process with their insights on needs in each area the project touches. The TCO process is outlined in a number of publications, and here are some elements of special emphasis in package development.

Operating costs. If innovation brings in new processes and equipment, the cost of training can increase. In food packaging, new aseptic and retort bring greater emphasis on operator training.

Quality control. More sophisticated processes can also bring more complex inspection processes. One example we saw was early in the military's meals-ready-to-eat (MRE) program. It was innovation with new equipment and new quality control steps compared to traditional canning. Ultimately, the process added a training burden to meet inspection requirements. Often such implications are inherent to new technology.

Design. If there is one place to focus on total costs it is in the design process; especially be alert for disconnects between marketing, packaging design and operations. If these groups aren't collaborating, it can have major cost implications. Examples include container shapes that slow down equipment operating speeds or those that can add damage or scrap to the process. Those factors add to the total cost of ownership and can be avoided if design and operations work together.

WHEN THE PACKAGING IS THE INNOVATION

Packaging and product innovation often can't be separated, and packaging's role is usually to support and enhance product innovation. Yet, in some cases, the package is the innovation—think of the Clorox pen with bleach for stain removal. The package has a larger role—it further differentiates and offers a marketplace edge. In this case, the packaging created a whole new category and made the innovator the category leader. But innovation that is led by packaging needs to be sold to corporate decision makers, and here's a way of doing it that is both low-cost and powerful.

First, *develop a list of packaging methods that could apply to your categories.* Think in terms of platforms that go beyond the traditional packaging in your category. For example, the single-serving cups you see for foodservice salad dressings. Could they hold a household cleaner to add convenience, performance or portability to answer consumer needs? Consider new technologies and new appliance-driven packaging options and new microwave solutions.

Use existing insights and criteria to build the case. Look at the competitive information you have, or analyze unmet consumer needs. Other factors include ownability, costs, SWOT information and testing. Build the list into a simple, "Basis for Interest" document; it is essentially a grid that organizes information to include a project title, and a summary of consumer benefits and need state connections. It also includes a description of when and where would we do this. It lists costs, and answers the questions, "Why would we do this?", "How would we do this?" and "How do we measure success?"

Set milestones and critical success factors. Build in the opportunity to get back to management with progress reports. Be sure those added contacts give management new information to keep up their interest. Since most new products have high fail rates, it is important to really think through how packaging can be a key enabler to help you be more successful.

Beware the pitfalls. If you present a solution to your management, be ready to implement it. That includes a business plan if it were brought to market. If you get senior management's attention, be able to deliver. If you don't, the tool loses its impact in your organization. Make sure you have support from collaborators within your organization. Do your operations people see a way to implement the process? Or, have you found a contract packager who can handle the process? In looking for pitfalls, also look for "Plans B, C, and D." They answer questions such as "If we have a delay in getting materials, can marketing delay the launch timetable?"

Here's another quick tool. Dredge up past successes, failures and discarded ideas. Ask what have you tried previously that has worked and look for the elements that made the project work. Look at what have you tried previously that did not work. Dredging through discarded ideas can be a fertile source of innovation.

Business and market conditions may have changed enough so what didn't work then may work now. Consider, too, that processes from the past may have application in emerging markets; look at what those markets need and match the character of processes from the past. You might get a very nice fit.

WHERE DO WE GO FROM HERE?

If we look at economic and business forecasts for individual countries and for the world, one thing is clear. We're in an era where linear extrapolation just doesn't work anymore. As an idea moves through the process from concept to a new product and package on the market, the process isn't going to be done the way it was done before. Success will go to those who have ability to truly interpret consumer needs, apply the right technology and do it in a way that gives an economic edge.

As we go forward, packaging becomes an integral part of that process. It has to be part of the process early, it has to support business strategies, and it has to involve the entire integrated value web to deliver what people, society and businesses need. That's true for the entire world. For those of you who are packaging professionals, incorporate this thinking in your work. For those senior managers who make the strategic decisions, know that packaging is a partner in the process and can be a catalyst to help your entire organization succeed.

Key action points for innovation and strategy

Innovation is a core business function

Packaging Action: Know that its primary goal is to deliver a business edge. Best practice companies do that by focusing on the consumer or customer and delivering better products to them.

Know your organization's tolerance for risk

Packaging Action: Before you can even start an innovation project, your have to know your organization's tolerance of risk and the roadblocks you can encounter. If the risk tolerance and an innovation project's risks aren't compatible, the project probably won't work.

Innovation is about focus

Gaining the right focus usually requires that the innovation's potential fit a company's business objectives.

Packaging Action: What areas do you want to investigate, what opportunity platforms work for your organization and what are your objectives? Have these in place to give you a manageable target.

Trust among collaborators and agreement on intellectual property are keys to successful open innovation.

Packaging Action: Before you start, negotiations need to "nail down" each side's limits on intellectual property. If you don't have that early, you run the risk of spending time and effort and then having the project terminate because the parties can't agree on who owns what.

Open innovation can be complex. It means coordinating with different functions across at least two organizations.

Packaging Action: Know the staff time that will go into a project and confirm that its payback can justify the added effort.

Have the right leader in place. It's all about leadership.

Packaging Action: Yes, tenacity is a key characteristic you need in a good leader—but not to the point that it results in rigidity.

References and Added Resources

CHAPTER 1. FIVE SECONDS THAT SIGNAL YOUR FUTURE

References

Hangzhou Wahaha Group Co., Ltd. en.wahaha.com.cn/

International Monetary Fund. *World Economic Outlook.* Sept. 2011.

The Hunt Group Inc., *Top 100 Global Packaged Goods Companies.* Non-food/beverage. 2008. www.huntsearch.com/assets/pdf/HGI%20Top%20100%20Global%20CPG%20Companies.pdf

Resources

The packaging value formula in practice

The packaging value formula's usefulness lies in comparing alternatives to define packaging's value contribution. Here's a simple example: sandpaper in a hardware store. The retailer sells single sheets of sandpaper from a bulk tray at $0.79 each, and a five-pack of sandpaper at $0.89 per sheet. In this example, the product doesn't change, so we hold its value constant. The package adds value, and the consumer experience improves with a neater workshop. We've added some estimated numbers to deliver a comparison between the options. We believe the package itself adds value, and we indicate that with a higher value. And, the package also improves the consumer's experience with the product being easier to find and identify in the home workshop.

$$\underset{\text{Single sheets}}{\text{Value } (2.53)} = \frac{\text{Product } (1) + \text{Package } (0) + \text{Experience } (1)}{\text{Price per sheet } (.79)}$$

$$\underset{\text{Five-pack}}{\text{Value } (2.58)} = \frac{\text{Product } (1) + \text{Package } (.2) + \text{Experience } (1.1)}{\text{Price per sheet } (.89)}$$

The formulas give us relative values to compare. They suggest that packaged sandpaper, at a higher price, can be competitive to loose sandpaper, a conclusion borne out by the hardware store's sales data.

CHAPTER 2. LOOK INSIDE THE CONSUMER'S BRAIN

References

Butschli, Jim. *Social media: A communications revolution causes a packaging revolution.* Packaging World magazine. Sept. 2011.

Penenberg. Adam L. *NeuroFocus Uses Neuromarketing To Hack Your Brain.* Fast Company.com. Aug. 8, 2011.

Pradeep, A. K. *The Buying Brain.* John Wiley & Sons, Inc. 2010.

CHAPTER 3. OPPORTUNITY IN A PERFECT RETAIL STORM

References

Best Global Brands 2011. Interbrand. www.interbrand.com/en/best-global-brands/best-global-brands-2008/best-global-brands-2011.aspx. 2011.

Bieler, Anne. *Retail packaging: Getting ahead of the curve.* A report from Packaging & Technology Integrated Solutions LLC. 2011.

Clifford, Stephanie. *Packaging Is All the Rage, and Not in a Good Way.* New York Times. Sept. 7, 2010.

Global Private Label Report: The Rise of the Value-Conscious Shopper. Nielsen Wire. http://blog.nielsen.com/nielsenwire/consumer/global-private-label-report-the-rise-of-the-value-conscious-shopper. March 2011.

100 Alternative Vending Machines. Trendhunter Lifestyle. http://www.trendhunter.com/slideshow/alternative-vending-machines#79

Pinto, Jim. *The Well Curve (Inverted bell curve).* http://www.jimpinto.com/writings/wellcurve.html. 2003.

Retail Packaging Defined. ECR UK. http://www.igd.com/index.asp?id=1&fid=5&sid=43&tid=58&cid=521.

Shulz, David P. *Top 100 Retailers.* NRF Stores. http://www.stores.org/STORES%2520Magazine%2520July%25202011/top-100-retailers. 2011.

Store brands cap a decade of growth with sales increases across the board. Private Label Manufacturers Association. http://plma.com/share/press/FOR_IMMEDIATE_RELEASE/New_Gains_Cap_Decade_of_SB_Growth_-_PLMA_2011_Yearbook.pdf. 2011.

Switching Channels: Global Powers of Retailing. Deloitte, LLP. 2012.

Tesco Homeplus Subway Virtual Store. YouTube. http://www.youtube.com/watch?v=nJVoYsBym88. 2011.

Wal-Mart And Target: Comparing Financials. Seeking Alpha. http://seekingalpha.com/article/312234-wal-mart-and-target-comparing-financials. Dec. 6, 2011.

CHAPTER 4. ANSWERS ARE IN THE VALUE WEB AND TEAMWORK

References

J. Peters Associates. *Contract Packaging: Strategic Opportunities & Profit Potential.* Packaging Strategies, a unit of BNP Media. 2005.

Kraft Foods Maps its Total Environmental Footprint. Kraft Foods. www.kraftfoodscompany.com/mediacenter/country-press-releases/us/2011/multi_media_12142011.aspx. 2011.

Peter Diamandis. *The future is brighter than you think.* CNN. http://www.cnn.com/2012/05/06/opinion/diamandis-abundance-innovation/index.html. 2012.

Resources

The scope of packaging

Often, we really don't see the reach of packaging in an organization. As an example of just how far it extends, here's a list of key functions where packaging becomes involved. It is based on the list of Packaging Technology Integrated Solutions capabilities.

Research identifies market needs, customer preferences, trends, new supplier capabilities, the regulatory environment, and production realities. It identifies relevant package design criteria that align with the organization's marketing requirements.

Design captures all relevant requirements, explores alternatives, and refines the strongest ideas; resulting in package designs that support business goals, marketing needs, promotion objectives, branding, safety issues, and portioning requirements. It answers the need for physical protection.

Graphics Management supports the highly specialized needs of graphic arts to ensure that all graphic assets are reproduced accurately, despite unique presses, exotic materials, unusual package shapes, and complex conversion and assembly steps.

Package Engineering addresses the physical aspects of the package, the materials used, production method, assembly steps, distribution methods and use—to ensure that all performance requirements are met. This includes determining optimal inner pack, case and pallet configurations.

Intellectual Property Management creates a competitive advantage by protecting and leveraging innovation while ensuring that the intellectual property of partners is treated appropriately.

Material Development is the innovation of new materials or processes to improve packages, meet performance goals, ensure safety throughout the life cycle, or achieve a unique competitive advantage.

Life-cycle Development is tied to brand management and marketing and uses research to identify opportunities to revitalize entire portfolios of packages. By planning efficient ways to innovate and replace your retiring packages, you can lower costs and protect the future viability of your brand.

Specification Management ensures that cross-functional input is obtained, all packaging elements are fully defined, and that these details are used to support better packaging decisions, consolidate raw material purchases, and manage supply-chain partners.

Sustainability translates sustainability goals into results by influencing design, engineering, material science, specification management, and procurement decisions. By creating holistic sustainability strategies, it is easier to develop packages that are environmentally friendly at every point in their life cycle.

Program Management includes orchestrating all the individual projects that relate to packaging and the packaging supply chain—ranging from one-time initiatives to ongoing programs.

Complex programs must tie in the full array of promotional elements that support all tactics and events.

Quality Assurance protects your brand by defining quality standards that are meaningful and measurable. Audits ensure that design specifications are produced and delivered to the consumers.

Analytics identifies meaning from performance as well as supplier and consumer data. Analytics also provide insights about ways to drive better business decisions, giving management a clear picture of how and where the spending is being directed.

Procurement Operations includes both strategic sourcing and supply-base management. That means guiding the activities necessary to find and manage the most appropriate supply-chain partners on an ongoing basis.

Planning and Optimization includes forecasting and predicting demand as well as providing those predictions to supply-chain partners to drive better business decisions (sourcing, material procurement, production planning, etc.) and ensure uninterrupted supply while minimizing inventory costs.

Education and Training helps people in your organization understand packaging issues, see emerging trends, build new skills and stay current with new regulations.

Assessing costs for contract packaging

Here's a way to assess the costs for contract packaging to see if they are acceptable and whether it makes more sense to set up the packaging operation in-house. It was published in the report *Contract Packaging: Strategic Opportunities & Profit Potential* authored by J. Peters Associates and distributed by BNP Media. Here are some of the business risks that impact the contract packaging decision:

- Does the new package differ significantly from existing packaging? Will the product/package fit on current packaging lines? Would you trade off product/package benefits to fit into existing packaging capabilities?
- ROI/payback hurdles: How big is the investment and what is the likelihood of reaching return/payback goals?
- Maturity of product category: Will the maturity of the product category affect potential success? Does it affect the risk of

failure, either with the risks of a new category, or the risks that competitors may respond in an established category?

- Estimated product lifespan: This factor may be particularly significant for line extensions or for promotional offerings.
- Need for confidentiality: Contract packagers sign confidentiality agreements, but breaches in confidentiality do occur. New developments have been compromised when a prospective client (and a competitor on the new development) looked behind a curtain while doing facilities evaluation. If the need for confidentiality is very high, a contract packager may not be the right answer.
- Product/package technology: How big a jump is it for all the participants in the packaging value chain? What is the likelihood that the technology can be implemented without time setbacks and process modifications?

The other step is calculating the break over point for contract packaging versus creating the packaging line in your own facility.

Contract Packaging Cost Analysis

Line	Cost comparisons	Co Packer	Owners Plant	Difference
1	Transportation inbound for raw material	$800	$1,250	–$450
2	Transportation out for finished product			$0.00
3	Packaging material	$6,550	$8,750	–$2,200
4	Warehousing cost	$350	$500	–$150
5	Transportation outbound	$900	$1,200	–$300
6	Waste and damage	$300.00	$200.00	$100.00
7	Line cost	$3,200	$5,000	–$1,800
8	Product change over cost @			
9	Package change over cost @			
10	Waste	$150	$100	$50
11	Over head & Profit	$1,500	$2,500	–$1,000
12	Other			
13	Totals	$13,750	$19,500	–$5,750
14	Line speed	120/min	150/min	
15	Total number of units produced	50,400	63,000	12,600
16	Cost per unit	$0.27	$0.31	–$0.04

Line 1—Freight costs may differ for different locations. Material should be the same cost assuming FOB pricing.

Line 2—If the product is made at the CPG's location and shipped in bulk to a co-packer, add in the cost of freight, totes, and handling. If this is the procedure, there is no difference in freight costs in line 1.

Line 3—Material plus freight charges.

Line 4—Includes raw materials, packaging, and finished goods.

Line 5—Transportation to distribution warehouse.

Line 6—Raw material and packaging, usually percentage of finished goods.

Line 7—Total cost to run each line for 8 hrs.

Line 8—This could be a fixed cost or lost time during the shift, which would reduce your output.

Line 9—Same as line 8, but just for a package change over.

Line 10—Lost units at packaging line and in storage x cost per unit.

Vested Outsourcing

This hybrid business model suggests collaboration in a way that fosters trust and mutual accountability for achieving results. Kate Vitasek is the proponent. Get her perspective from The University of Tennessee Center for Executive Education. www.vestedoutsourcing.com/

CHAPTER 5. DESIGN: A SOUL ISSUE THAT HAS TO BE HOLISTIC

References

Sabena, Patricia. *Herbal Essences: Contemporizing Brand Equity.* 2005.

The Packaging Design Brief in Detail

This is the critical document to shape packaging design. Below are the major groupings and details for a packaging design brief.

1. Design objectives
    - Describe what situation instigated the project. Is this a new product or existing redesign?
    - Define design objectives and measurement criteria

- Purchase intent
- Brand recall
- Brand visibility and key associations
- Must haves, wants and boundaries
- Form, count, sizes, preliminary claims

2. Brand positioning; understanding of equities
    - Marketing is the source for this information
    - What are the brand's key attribute statements?
    - Define the importance of sustainability and relationship to the brand
    - What are brand positioning elements and equities?
    - How will this change make the brand unique?
3. Consumer/Purchaser Information
    - Input from marketing, consumer insights and sales define the target consumer
    - Get other data such as demographics, psychographics, etc.
    - Where is the product purchased?
    - Buying habits
    - Sustainability interest
    - Insights on use: transport, storing, opening, dispensing, etc.
    - Any consumer or user absolutes?
    - Share relevant prior research results
4. Customer Information
    - Often comes from sales team
    - What are the sales channels?
    - How will packages be shelved, displayed?
    - Does a plan-o-gram exist?
    - Will this package be part of a display or special pack?
    - Account specific requirements or constraints
    - Customer category likes/dislikes
    - Current selling price and relative positioning in the category
    - The same package may be displayed many ways; see chapter on retail
5. Competitive products and creating a differentiated position
    - Who is your competition?
    - What are their strengths and threats?
    - Has the competitive product or packaging changed recently? If so, how?

- How does the existing package compare to the competitive offerings? (*See the Competitive retail audit in Chapter 3.*)

6. Product Parameters
   - What are the key product attributes and benefits?
   - Technical information on the product
   - Shelf life requirements
   - Environmental impacts of light, oxygen, moisture, etc. on the product
   - Describe the general processing and filling operation
7. Package Design Requirements
   - Are there existing package attributes to be maintained (color, texture, size, shape, etc.)?
   - Have any other formats been tried previously? If so, describe.
   - What are the secondary packaging requirements?
   - How is the product sold by channel?
   - What are the shipping modes and distances?
   - What is the target cost of goods sold (COGS) or return on investment (ROI)?
8. Sustainability considerations
   - The importance of eco-friendliness and "green" packaging has become paramount, in terms of cost and what consumers want: these issues have to be part of the design brief.
   - What are the key issues of concern? They vary by product category and even company. For some, the concern is cutting costs by minimizing some materials.
   - What are the product, brand and company sustainability objectives?
   - Here is a key question to ask: What is the package's anticipated end of life?
   - Has the cube been optimized?
   - Can recycled materials be used?
   - What sustainability certifications are available for use? The consumer research cited in the chapter on green issues points to value is using certifications. Where and how can you incorporate them into your package graphics?

9. Manufacturing requirements
   - Will this be packaged internally or via contract packing?
   - Machine filled or hand packed?
   - What kind of equipment is used?
   - What are the packaging speeds?
   - Date/lot codes required? (If so, are the location and requirements specified?)
10. Marketing, sales and procurement
    - What are the anticipated production volumes?
    - What are the strengths and weaknesses of existing graphics?
    - Will language translations be required?
    - Who is the anticipated packaging supplier?

What do colors mean?

This perspective on what colors mean comes from Jolly Hanspal, a senior designer at Boxer Creative. Boxer is headquartered in the United Kingdom and has a global perspective on design. Some color connotations have specific regional differences and you will probably want to get the input of a local design firm on specific meanings for a specific country.

White is associated with purity and freshness so is often used on bathroom and kitchen packaging to suggest a "clean" feeling. It is also used for dairy products by product association. White in the world of the supermarket represents value product and is used across the board in all supermarkets for this purpose. In contrast to this, white can also mean premium if used in the correct way—giving the feeling of "fresh" and "pure". The "fresh, pure & clean" feeling of white crosses over into health/beauty and "feel good" products also.

Green is associated with eco-friendly products. Green is used often on product banners to bring attention to the no-fat and low-fat labeling, conveying the impression of health and 'good for you.' Green gives the feeling of "natural" and "of the earth" which also crosses into health/beauty and "feel good" products.

Brown kraft paper and paperboard was once perceived as cheap, but now gives the strong perception of eco-friendly—more so than green.

Red is used to grab the consumer's attention. Coca-Cola, the world's most recognizable brand, has long been associated with red. Yet, caution needs to be taken because of connotations of being too aggressive.

Yellow is known as being one of the most visible packaging colors. It also has the ability to make products appear larger on the shelf.

Black projects luxury and premium quality. It is used to convey top quality, sophistication.

Gold, long linked to royalty and luxury, is now thought to be associated with "cheap" in the shopper's mind. It is often used on inexpensive or imitation generic packaging. On the flipside, if used correctly in application it can still purvey premium. Soft gold, however, is used often on children's products because it is considered a happy, sunny color.

THE COLOR MARKETING GROUP is an organization that focuses on color trends. On the Internet, search for Color Marketing Group.

CHAPTER 6. KNOW THAT GREEN IS NORMAL

References

As U.S. Bottled Water Sales Decline 5.2%, Nestlé Waters Taps Into Developing Markets. Food&waterwatch. 2011. www.foodandwaterwatch.org/pressreleases/as-u-s-bottled-water-sales-decline-5-2-nestle-waters-taps-into-developing-markets/

EcoFocus Worldwide. http://ecofocusworldwide.com/

Green branding & sustainable design. Landor. 2010. http://landor.com/#!/about/capabilities/green-branding-sustainable-design/

Shoppers Want Their Packaging "Green." Perception Research Services. 2012. http://www.prsresearch.com/fileUploads/ShoppersWant TheirPackagingGreen_4.pdf

The story behind Apple's environmental footprint. Apple Inc. 2012. http://www.apple.com/environment/

Whole Foods Market. *Green Mission Report.* 2012. Austin, TX. On-line at www.wholefoodsmarket.com/pdfs/2012GreenMissionReport.pdf.

Environmental impact of packaging

Lars Erlöv, Cathrine Löfgren, Anders Sörås. *Report No. 194—Packag-*

ing—a Tool for the Prevention of Environmental Impact. Stockholm: STFI-Packforsk, June 2000.

eBay's sustainable packaging experience

In assessing the eBay experience with the Value formula, we see how packaging alone cannot offset the strategic benefit lost through the experience of entering data into a less-than-average tracking system. The first value formula is the baseline, assuming a "1" value for all components. The second value formula shows our estimates of factors that changed the value for the package, along with a cost increase, and a lower experience factor that does not deliver the desired experience. The result is that the new box has less estimated value than the older container, but for a reason unrelated to the packaging itself.

$$\text{Value (3.0)} = \frac{\text{Product (1)} + \text{Package (1)} + \text{Experience (1)}}{\text{cost per box (1)}}$$

$$\text{Value (2.89)} = \frac{\text{Product (1)} + \text{Package (1.1)} + \text{Experience (.85)}}{\text{cost per box (1.02)}}$$

In this formula, the product is the eBay shipment, and the package is the reusable box. We give the box a 1.1 for its perceived sustainability benefit, a slight benefit. However, the experience factor—in this case entering tracking data—is below optimum. The cost per box is the upcharge for graphics and structure minus reduced requirements for tape.

This is for illustration only, and numbers that differ from our estimate would yield a different value.

CHAPTER 7. GLOBAL GROWTH IS FOR THE TAKING, WITH CAUTION

References

Emerging vs. developed countries' GDP growth rates 1986 to 2015. Real-World Economics Review Blog. 2011. http://rwer.wordpress.com/2011/03/24/emerging-vs-developed-countries-gdp-growth-rates-1986-to-2015/

Focus on growth markets. Goldman Sachs. 2012. www.goldmansachs.com/our-thinking/focus-on/growth-markets/index.html?cid=PS_01_20_06_99_01_03_t1

History—the growth history of Wahaha. Wahaha Fulin Co. 2012. http://en.wahaha.com.cn/aboutus/history/index2.shtml

O'Neil, Shannon K. *The Bright Side of the Global Economy: The Middle Class Is Growing.* The Atlantic. Nov. 9, 2011.

Population Projection Tables by Country and Group. The World Bank Group. 2011. web.worldbank.org/WBSITE/EXTERNAL/TOPICS/EXTHEALTHNUTRITIONANDPOPULATION/EXTDATASTATISTICSHNP/EXTHNPSTATS/0,,contentMDK:21737699~menuPK:3385623~pagePK:64168445~piPK:64168309~theSitePK:3237118~isCURL:Y,00.html

Prahalad, C.K. *The Fortune at the Bottom of the Pyramid.* 2004.

Rangnekar, Amit, Ph.D. *Chic-Shampoo Sachet.* 2007 http://www.scribd.com/doc/2302947/Case-Chic-Shampoo-Rural-Revolution

World Economic Outlook. International Monetary Fund. 2011.

World Economic Outlook. International Monetary Fund. 2012.

World Gross Domestic Product (GDP) by region. U.S. Energy Information Administration. 2011. www.eia.gov/oiaf/aeo/tablebrowser/#release=IEO2011&subject=0-IEO2011&table=3-IEO2011®ion=0-0&cases=Reference-0504a_1630

World Economic Forum on Africa 2012. World Economic Forum. 2012. www.weforum.org/events/world-economic-forum-africa-2012

CHAPTER 8. IT'S A RISKY WORLD OUT THERE

References

Anti-Counterfeiting and Product Protection Program (A-CAPPP). Michigan State University. www.a-cappp.msu.edu

ISO 31000—Risk Management. International Organization for Standardization. 2009. (Anticipate ISO 31004—*Risk management—Guidance for the implementation of ISO 31000* in 2013.)

Risk Assessment Matrix (RAM) Process. Michigan State University School of Criminal Justice.

Spink, J. Overview of the Selection of Strategic Authentication and Tracing Programmes. 2012. In: Albert Wertheimer and Perry Wang (Eds.), *Counterfeit Medicines: Volume I. Policy, Economics, and*

Countermeasures. Hertfordshire, United Kingdom: ILM Publications, 111-128, ISBN 9781906799083.

The ACTA Group. *TSCA Reform—Business Strategies in Times of Political Gridlock*. 2012. Environmental Expert.com. www.environmental-expert.com/articles/tsca-reform-business-strategies-in-times-of-political-gridlock-284793/view-comments

Resources

Risk register tool

The "risk register" is a tool suggested to us by Bob Pojasek, Ph.D. He leads the sustainability consulting practice at The Shaw Group. The tool is in the form of a database arrayed here:

Ref. No.	Description	Inherent risk level (impact/ probability)	Description, evaluation of current control and mitigation methods	Residual risk level (impact/ probability)	Acceptable Risk? (impact/ probability)	Action

Keys to the process are the column "inherent risk"; it is assigned in conjunction with a corporate risk management professional. The residual risk value is assigned after steps to mitigate it have been taken. The next question is "acceptable" or "unacceptable," and those falling in the "unacceptable" category have to be dealt with.

Risk assessment grids

We showed one version of a risk assessment map in Chapter 8. A number of other variations exist, and most use a threat-probability axis such as the one that follows. Basic risk management tries to identify unacceptable risks, which is shown in the upper right hand corner in this version of a risk assessment grid above the black line. Risk management says to manage risks to bring them to an acceptable level (below and left of the line) and insure against the unacceptable risks.

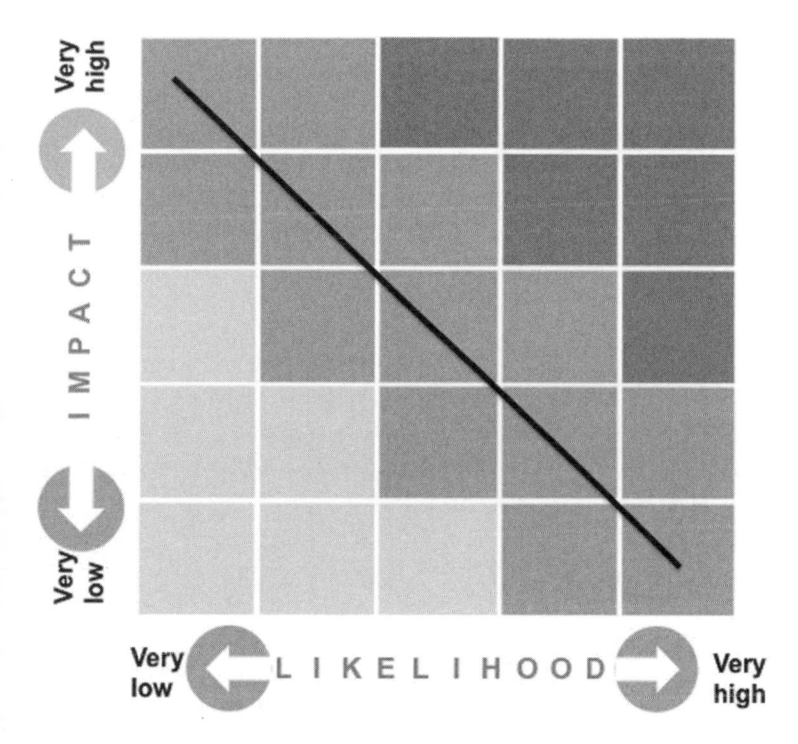

Anti-fraud techniques

This list comes from John Spink, Ph.D., Associate Director of the Anti-Counterfeit and Product Protection Program at Michigan State University. It focuses only on packaging technologies, and other components comprise a holistic anti-fraud program. For detailed references on anti-fraud methods, go to http://www.a-cappp.msu.edu/users/dr-john-spink

Technique	Tactics
Substrates and components	Security paper, films, security threads
Optically variable devices	Holograms, reflective devices, color shifting inks, films and others
Print methods	Designs, security print features, digital watermarks
Coatings, inks	Security inks and additives, taggants
Coding	Serialization, fingerprinting

CHAPTER 9. STRATEGIC THINKING, INNOVATION: PACKAGING'S PIVOT POINTS

References

Innovate your way to world class packaging. Packaging Technology Integrated Solutions, a division of HAVI Global Solutions. 2012. A presentation at the 6th Annual Food Technology & Innovation Forum.

Packaging Innovation: A key business tool and enabler for growth and profitability. Packaging & Technology Integrated Solutions LLC. 2009. Published by Packaging Strategies.

Stage-Gate International. www.stage-gate.com

Top 10 Lessons on the New Business of Innovation. MIT Sloan Management Review. Winter 2011.

Stage-Gate process

The Stage-Gate innovation process is a product of Stage-Gate international. In its basic form, it sets parameters for at least five stages in an innovation project—concept development, feasibility, commercialization, tracking and follow up. After each stage is a "gate," the set of parameters that can be used to measure if the stage meets project criteria.

Our belief is that management processes of this type offer several advantages. They:

- Encourage solid, up-front homework and drives an understanding of the problem. It helps to incorporate related work such as relevant store audits, etc.
- Emphasize voice of the customer/consumer
- Promote a product advantage, something that delivers value for the user
- Set up an early and clear package and product definition
- Keep project "honest" with tough go/kill decisions
- Emphasize accountable cross-functional teams

About the Authors

JIM PETERS is an author, editor and consultant whose experience in packaging goes back 40 years. He has written about packaging milestones such as adoption of the Universal Product Code, the Tylenol event, the advent of sustainability, and packaging's growth as a marketing tool and as a strategic business function. He's been the editor of Packaging Engineering and BrandPackaging magazines. He's written and edited reports on contract packaging, retail packaging and global supply chain developments; he has been a speaker at events such as the Brazilian Packaging Congress, Packaging Association of Canada meetings and marketing conferences. He's been director of education for the Institute of Packaging Professionals, and is a Fellow of the organization. Peters' degree in journalism is from Northern Illinois University.

BRIAN HIGGINS is Senior Vice President of Consulting & Business Solutions at HAVI Global Solutions (HGS). He leads the consulting business and he works with customers and suppliers to create competitive advantages through packaging. Brian's emphasis is on strategic directions across the global supply chain. He has deep supply chain management experience as both a practitioner and a management consultant. Higgins has focused experience within the retail, consumer goods and high technology sectors. Prior to joining HGS, Brian was the director of the Supply Chain Strategy practice for KPMG Consulting. He started his career as a business process reengineering manager for a division of Standard Register. Brian holds an International MBA from the University of South Carolina and Wirtshaftsuniversitat Wien (Vienna University of Business and Economics). He completed his undergraduate studies in London, England, at the London School of Economics and at the University of Dayton (Ohio).

MICHAEL RICHMOND, Ph.D., is Vice President, Packaging Technology Integrated Solutions (PTIS), a division of HAVI Global Solutions. He is a strategic business and technical leader with 25 years of experience across PTIS, Kellogg, Kraft, and Michigan State University. Mike was instrumental in developing and implementing strategic plans and programs nationally and globally for both Kraft and Kellogg. He led the research and development component of strategic sourcing at both companies. Mike also brings a consumer and trade focus to the development of solutions to packaging problems globally. Mike has completed executive programs at both Harvard and Thunderbird. Mike's Ph.D. in food science is from Michigan State University. He entered the Packaging Hall of Fame in 2011. Richmond continues to support academics through the PTIS Packaging Endowment at Michigan State University and lectures at both Michigan State and Western Michigan University in packaging, food marketing and MBA programs.

Those who added their insights to this book

The holistic thinking in this book isn't the product of one person's mind. And it probably couldn't be. Here are the people who contributed ideas and content to this book.

Jonathan Asher
Vice President
Perception Research Services

Shane Bertsch
Vice President, Global Packaging
HAVI Global Solutions

Anne Bieler
Senior Associate
Packaging Technology Integrated Solutions
HAVI Global Solutions

Peter Borowski
Senior Design Director
Kraft Foods

Tim Brown
Manager Packaging Consulting Services
Packaging Technology Integrated Solution
HAVI Global Solutions

Todd Bukowski
Senior Associate
Packaging Technology Integrated Solutions
HAVI Global Solutions

Paul Castledine
Chairman & Chief Creative Officer
Boxer Creative

Marc Campbell
Senior Director of Strategic Sourcing
HAVI Global Solutions

Neil Darin
Senior Program Manager, Innovation and Sustainability
HAVI Global Solutions

Linda Gilbert
CEO
EcoFocus Worldwide

Jack Gordon
President
AccuPOLL Research

Alan Isacson
CEO
ABI Marketing Public Relations

James B. Kauffman
President
The Everest Group

Ross Lee
Senior Consultant
Packaging Technologies Integrated Solutions
Havi Global Solutions

John Mahaffie
Co-founder
Leading Futurists LLC

Phil McKiernan
Vice President
Packaging Technology Integrated Solutions
HAVI Global Solutions

Bob Pojasek
The Shaw Group

Patrick Rodmell
Principal
Rodmell Associates

John Spink, Ph.D.
Associate Director and Assistant Professor
Anti-Counterfeiting and Product Protection Program
School of Criminal Justice
Michigan State University

Scott Young
President
Perception Research Services

Brian Wagner
Vice President
Packaging Technology Integrated Solutions
HAVI Global Solutions

Index